Divided Solids Mechanics

There are no such things as applied sciences,
only applications of science.
Louis Pasteur (11 September 1871)

Dedicated to my wife, Anne, without whose unwavering support, none of this
would have been possible.

Industrial Equipment for Chemical Engineering Set

coordinated by
Jean-Paul Duroudier

Divided Solids Mechanics

Jean-Paul Duroudier

ELSEVIER

First published 2016 in Great Britain and the United States by ISTE Press Ltd and Elsevier Ltd

ISTE Press Ltd
27-37 St George's Road
London SW19 4EU
UK

www.iste.co.uk

Elsevier Ltd
The Boulevard, Langford Lane
Kidlington, Oxford, OX5 1GB
UK

www.elsevier.com

Notices

For information on all our publications visit our website at http://store.elsevier.com/

British Library Cataloguing-in-Publication Data
A CIP record for this book is available from the British Library
Library of Congress Cataloging in Publication Data
A catalog record for this book is available from the Library of Congress
ISBN 978-1-78548-187-1

Printed and bound in the UK and US

Contents

Preface

The observation is often made that, in creating a chemical installation, the time spent on the recipient where the reaction takes place (the reactor) accounts for no more than 5% of the total time spent on the project. This series of books deals with the remaining 95% (with the exception of oil-fired furnaces).

It is conceivable that humans will never understand all the truths of the world. What is certain, though, is that we can and indeed must understand what we and other humans have done and created, and, in particular, the tools we have designed.

Even two thousand years ago, the saying existed: "faber fit fabricando", which, loosely translated, means: "*c'est en forgeant que l'on devient forgeron*" (a popular French adage: *one becomes a smith by smithing*), or, still more freely translated into English, "practice makes perfect". The "artisan" (faber) of the 21st Century is really the engineer who devises or describes models of thought. It is precisely that which this series of books investigates, the author having long combined industrial practice and reflection about world research.

Scientific and technical research in the 20th century was characterized by a veritable explosion of results. Undeniably, some of the techniques discussed herein date back a very long way (for instance, the mixture of water and ethanol has been being distilled for over a millennium). Today, though, computers are needed to simulate the operation of the atmospheric distillation column of an oil refinery. The laws used may be simple statistical correlations but, sometimes, simple reasoning is enough to account for a phenomenon.

Since our very beginnings on this planet, humans have had to deal with the four primordial "elements" as they were known in the ancient world: earth, water, air and fire (and a fifth: aether). Today, we speak of gases, liquids, minerals and vegetables, and finally energy.

The unit operation expressing the behavior of matter are described in thirteen volumes.

It would be pointless, as popular wisdom has it, to try to "reinvent the wheel" – i.e. go through prior results. Indeed, we well know that all human reflection is based on memory, and it has been said for centuries that every generation is standing on the shoulders of the previous one.

Therefore, exploiting numerous references taken from all over the world, this series of books describes the operation, the advantages, the drawbacks and, especially, the choices needing to be made for the various pieces of equipment used in tens of elementary operations in industry. It presents simple calculations but also sophisticated logics which will help businesses avoid lengthy and costly testing and trial-and-error.

Herein, readers will find the methods needed for the understanding the machinery, even if, sometimes, we must not shy away from complicated calculations. Fortunately, engineers are trained in computer science, and highly-accurate machines are available on the market, which enables the operator or designer to, themselves, build the programs they need. Indeed, we have to be careful in using commercial programs with obscure internal logic which are not necessarily well suited to the problem at hand.

The copies of all the publications used in this book were provided by the *Institut National d'Information Scientifique et Technique* at Vandœuvre-lès-Nancy.

The books published in France can be consulted at the *Bibliothèque Nationale de France*; those from elsewhere are available at the British Library in London.

In the in-chapter bibliographies, the name of the author is specified so as to give each researcher his/her due. By consulting these works, readers may gain more in-depth knowledge about each subject if he/she so desires. In a reflection of today's multilingual world, the references to which this series points are in German, French and English.

The problems of optimization of costs have not been touched upon. However, when armed with a good knowledge of the devices' operating parameters, there is no problem with using the method of steepest descent so as to minimize the sum of the investment and operating expenditure.

1

Mechanical Characteristics of Divided Solids

1.1. Two simple properties

1.1.1. The size of particles

In the following, we will use the abbreviation D.S. for divided solid.

In terms of particle size alone, we distinguish:

– *powdered* solids, that is powders consisting of:

- ultrafine ($d_p < 20$ μm) talc, flour, some pigments and coloring agents. These products are a result of fine grindings;

- fine (20 μm $< d_p < 100$ μm);

– *granules* consisting of artificial granules or natural particles such that:

$$d_p \geq 500 \text{ μm}$$

These particles are called "granules" as opposed to "fine";

– *intermediate* solids often as a result of current grindings or usual crystallizations in a mother liquor:

$$100 \text{ μm} < d_p < 500 \text{μm}$$

The expression "granular solid" is to be avoided.

In [WIE 75], Wiegbardt reviews the different properties of D.S. as well as the measurement of these latter.

1.1.2. Compressibility

If the diameter of the powdered solids is inferior to 60 or 100 μm, van der Waals forces of attraction are predominant compared to forces of gravity.

If the porosity drops locally below a certain value, due to interactions between particles, bonds are formed, which lead to clusters.

If we try to fluidize a powder, the gas will flow in preferential passages separated by high solid density regions.

If we want to transport a powder pneumatically and in a dense phase, any local drop in the gas velocity will lead to the formation of a solid plug.

During the emptying of a hopper, the flow of powder takes a pulsed behavior, because a local increase in solid density is sufficient to lead to a solid plug, which almost provokes the stop of the flow (arching, doming).

In order to define this property of powders, some authors suggest we use Hausner's index, which is defined as the ratio between the apparent mass density ρ_a of the solid after compression and density in the "aerated" state, that is loose or movable:

$$I_H = \frac{\rho_a \text{ "after compression"}}{\rho_a \text{ "aerated"}} \quad (I_H > 1)$$

The aerated mass density ρ_a is explained in Appendix 1 for several D.S.

Ergun [ERG 51] provides a method to determine the true (intrinsic) mass density of a D.S:

$$\rho_s = \frac{\rho_a}{I - \varepsilon}$$

The compression is obtained by repeated shocks (tapping) and the "aerated" state is that of the powder collected in a cylindrical container located under a vibrated sieve such that the flow into the container is uniform on the area.

The shocks are supposed to simulate the ease of appearance of a strong solid density during a flow.

In their publication, Harnby *et al.* [HAR 87] specify the use of this concept using an example. However, in flow studies, Hausner's index is not convenient and Jenike suggests that:

$$-\frac{dLnV}{d\,LnP} = \frac{d\,Ln\rho_a}{dLn\sigma_{\ell c}} = \beta = \text{"coefficient of compressibility"}$$

where $\sigma_{\ell c}$ is the major principal contact stress. The contact stresses only act on the solid skeleton.

Remember that if we consider a D.S. as a continuous medium, the tensor matrix of the stresses can be written as:

$$[\Sigma] = \varepsilon P[I] + (1 - \varepsilon)[\Sigma_2]$$

P: pressure of the interstitial fluid (Pa)

[I]: unit matrix (matrix of pressure stresses)

σ: stress in the continuous medium (Pa)

ε: porosity of the medium

[Σ]: matrix of stresses

[Σ_2]: matrix of the contact stresses.

The compressibility β is different depending on the products and whether the product is immobile or flowing.

Type of product:	β
Hard solid (steel balls)	0
Inorganic solid	< 0.05
Spongy solid	> 0.1
Fibrous solid	#1

NOTE.–

We sometimes exert pressure on divided solids saturated with liquid to drive out the interstitial liquor. The permitted variation law is similar to the previous one:

$$\frac{V_o - V_m}{V - V_m} = Ln\left(\frac{P}{P^*} + e\right)$$

where V_o and V_m are the initial volume and the minimum volume (for $P \rightarrow \infty$). The parameter e is the base of natural logarithms and P^* is also a parameter that characterizes the compressibility of the product. The pressure P is measured in rel. bar.

NOTE.–

A D.S. can be:

– soft (loose) if it results from a reduced speed finish;

– compact if it has been subjected to pressure. This pressure is said to be of consolidation. To us, it will always be less than 50,000 Pa.

NOTE.–

The reader interested in the microscopic behavior of D.S. can refer to Cambou and Jean [CAM 01].

1.2. The mechanics of continuous media

1.2.1. Notion of stress in a solid

Consider a solid domain limited by an area S and let there be an element dS of this area on which the external medium exerts a force $\vec{T}$. We write:

$$\vec{T} = \vec{t} \ dS$$

where $\vec{t}$ is the ratio of a force to an area i.e. is a stress and we can split this vector in a component $\vec{\sigma}$ perpendicular to the surface dS and a component $\vec{\tau}$ tangent to the surface dS.

In the classical resistance of materials applied to metals, plastic materials and composite materials, the perpendicular component $\vec{\sigma}$ is considered positive if it's a traction, that is if $\vec{\sigma}$ is directed toward the exterior of the solid field. We say that the unit vector of the perpendicular stress is that of the *outgoing* perpendicular vector.

On the contrary, in soil mechanics and in the mechanics of divided solids, the opposite convention is used, the compressions are considered positive, that is throughout the *entering* normal vector $\vec{n}$.

Now let's consider a plane (P) defined at the point M by the stress $\vec{t}$ and the entering normal vector $\vec{n}$ and let ox be an axis of this plane.

We notice that the oriented angle β in this plane, which brings ox in the direction of $\vec{n}$ changes sign depending on the observer standing on one side or the other of the plane (P).

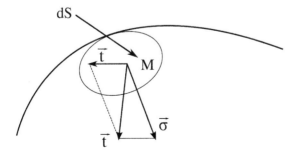

Figure 1.1. *Stress at the point M*

NOTE.–

At a given point, we can only split down the stresses into vector components if they are exerted on the same surface area. If it is not the case, we can only split forces into vector components, which is the product of stresses and the surface elements of different orientations.

1.2.2. Equilibrium of forces and tensor of the stresses in plane coordinates

Consider the right-angled triangle PAB in the xoy plane.

The unit vectors along ox and oy are $\vec{i}$ and $\vec{j}$.

t_x and t_y are the components along ox and oy of the stress $\vec{t}$ exerted on side AB of the triangle PAB.

$$\vec{t} = \vec{i}t_x + \vec{j}t_y$$

Let $\vec{n}$ be the unit vector of the entering normal vector to AB. The compression stress will be:

$$\vec{\sigma} = \vec{n}\sigma$$

Let β be the angle oriented in the anticlockwise direction (in the opposite direction to that of the hands of a clock) from ox and $\vec{n}$ (on the Figure 1.2, this angle is such that: $\pi < \beta < 3\ \pi/2$).

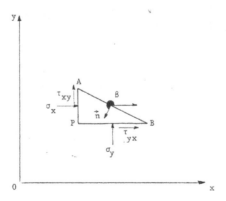

Figure 1.2. *Equilibrium of forces*

The force exerted by the stress $\vec{t}$ on the side AB of length ds is:

$$\vec{t}ds = ds(\vec{i}t_x + \vec{j}t_y)$$

The stresses normal to PA and PB, that is σ_x and σ_y are measured in the direction of the axes ox and oy, that is in the direction of entering normal vectors. The shearing stresses τ_{yx} and τ_{xy} respectively on PB and PA are also measured positively respectively in the direction of the axes ox and oy.

The force exerted on the side PA of positive length is proportional to:

$$ds\cos PAB = ds\cos(ox, -\vec{n}) = ds\cos[(ox, \vec{n}) - \pi] = -ds\cos\beta$$

$$-ds\cos\beta[\vec{j}\tau_{xy} + \vec{i}\sigma_x]$$

Similarly, along PB:

$$-ds\sin\beta[\vec{i}\tau_{yx} + \vec{j}\sigma_y]$$

Let's say the sums of the components of entering forces along ox and oy are nil (the "entering forces" are the forces exerted on the triangle from the exterior):

Along *ox* (vector $\vec{i}$): $t_x = \sigma_x \cos\beta + \tau_{yx} \sin\beta$

Along *oy* (vector $\vec{j}$): $t_y = \sigma_y \sin\beta + \tau_{xy} \cos\beta$

These two equalities can be rewritten in matrix form as:

$$\begin{bmatrix} t_x \\ t_y \end{bmatrix} = \begin{bmatrix} \sigma_x & \tau_{yx} \\ \tau_{xy} & \sigma_y \end{bmatrix} \cdot \begin{bmatrix} \cos\beta \\ \sin\beta \end{bmatrix}$$

Or:

$$(\vec{t}) = (\Sigma)(\vec{n})$$

(Σ) is the stress matrix.

1.2.3. *Equilibrium of forces of volume and of moments: stress tensor symmetry*

Let f_x and f_y be the forces of volume. The classic reasoning of the "small parallelepiped" (see Laroze [LAR 74]) leads to the following result:

For a tiny displacement dx and dy, the equilibrium of stress variations of the volume forces is written as:

$$\frac{\partial \tau_{xy}}{\partial y} + \frac{\partial \sigma_x}{\partial x} = f_x \qquad [1.1]$$

$$\frac{\partial \sigma_y}{\partial y} + \frac{\partial \tau_{yx}}{\partial x} = f_y \qquad [1.2]$$

(The vector $\vec{n}$ being entering, the signs of the stresses are modified.)

The variation of the resulting moment from these forces has to be nil:

$$x f_y - y f_x + \frac{\partial}{\partial x}\left(x \tau_{yx}\right) + \frac{\partial}{\partial y}\left(x\, \sigma_y\right) - \frac{\partial}{\partial y}\left(y \tau_{xy}\right) - \frac{\partial}{\partial x}\left(y\, \sigma_x\right) = 0$$

Replacing f_x and f_y by their values obtained from [1.1] and [1.2] and developing the derivations, we obtain:

$$\tau_{yx} - \tau_{xy} = 0$$

The stress tensor is therefore symmetrical:

$$\tau_{yx} = \tau_{xy} = \tau$$

1.2.4. *Principal directions and eigenvectors*

Let's find a direction $\vec{n}$ forming an angle β with ox such that the stress exerted on the perpendicular surface to $\vec{n}$ is proportional to $\vec{n}$:

$$\vec{T} = \sigma \vec{n}$$

$$\sigma_x \cos\beta + \tau \sin\beta = \sigma \cos\beta$$

$$\tau \cos\beta + \sigma_y \sin\beta = \sigma \sin\beta$$

Eliminating $\sin\beta/\cos\beta$ from these two equations, we obtain:

$$\sigma^2 - \left(\sigma_x + \sigma_y\right)\sigma - \tau^2 + \sigma_x\sigma_y = 0$$

$$\sigma_1 \text{ or } \sigma_2 = \frac{1}{2}\left(\sigma_x + \sigma_y\right) \pm \sqrt{\frac{1}{4}\left(\sigma_x - \sigma_y\right)^2 + \tau^2}$$

The principal directions β_1 and β_2 are then defined by dividing the two equations by $\cos\beta$:

$$\tan\beta_1 = \frac{\sigma_1 - \sigma_x}{\tau} = \frac{\tau}{\sigma_y - \sigma_1}$$

$$\tan\beta_2 = \frac{\sigma_2 - \sigma_x}{\tau} = \frac{\tau}{\sigma_y - \sigma_2}$$

This procedure for finding the principal directions would still be possible even if we had not made use of the symmetry of the stress tensor.

1.2.5. Mohr's circles in three dimensions [LAR 74]

In a direction $\vec{n}$ of space, we can exert a stress $\vec{t} = \Sigma.\vec{n}$ at a point P where Σ is the stress matrix. Let α, β, and γ be the components of $\vec{n}$ in a unique trirectangular base of Σ. Let's write the following equations, where σ and τ are the components of $\vec{t}$:

– in the direction $\vec{n}$: normal stress: σ

– in the perpendicular plane to $\vec{n}$: shear: τ

– Pythagoras' theorem:

$$\sigma_1^2\alpha^2 + \sigma_2^2\beta^2 + \sigma_3^2\gamma^2 = \sigma^2 + \tau^2$$

The magnitude of $\vec{n}$ is equal to 1:

$$\alpha^2 + \beta^2 + \gamma^2 = 1$$

The components of $\vec{t}$ are $\sigma_1\alpha$, $\sigma_2\beta$ and $\sigma_3\gamma$. Let's say that the sum of orthogonal projections of these components in the direction of $\vec{n}$ is equal to σ.

$$\sigma_1\alpha^2 + \sigma_2\beta^2 + \sigma_3\gamma^3 = \sigma$$

This system of three equations can be resolved in terms of α^2, β^2 and γ^2:

$$\alpha^2 = \frac{\tau^2 + (\sigma - \sigma_2)(\sigma - \sigma_3)}{(\sigma_1 - \sigma_2)(\sigma_1 - \sigma_3)}$$

$$\beta^2 = \frac{\tau^2 + (\sigma - \sigma_3)(\sigma - \sigma_1)}{(\sigma_2 - \sigma_3)(\sigma_2 - \sigma_1)}$$

$$\gamma^2 = \frac{\tau^2 + (\sigma - \sigma_1)(\sigma - \sigma_2)}{(\sigma_3 - \sigma_1)(\sigma_3 - \sigma_2)}$$

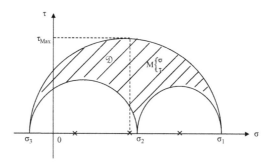

Figure 1.3. *Area of existence of the stress $\vec{t} = \vec{\sigma} + \vec{r}$ at the point M*

We can always assume that $\sigma_1 > \sigma_2 > \sigma_3$. Let's say $\alpha^2 > 0$, $\beta^2 > 0$ and $\gamma^2 > 0$, that is:

$$\sigma^2 + \tau^2 - (\sigma_2 + \sigma_3)\sigma + \sigma_2\sigma_3 \geq 0$$

$$\sigma^2 + \tau^2 - (\sigma_3 + \sigma_1)\sigma + \sigma_3\sigma_1 \leq 0$$

$$\sigma^2 + \tau^2 - (\sigma_1 + \sigma_2)\sigma + \sigma_1\sigma_2 \geq 0$$

The equalities are the equations of three circles in the coordinate system (σ, τ). The inequalities correspond to the hatched area where the representative point of the stress $\vec{t}$ must be found.

1.2.6. Two-dimensional case [LAR 74]

Mohr's three circles from the Figure 1.3 are reduced to only one circle in the case where the stress σ_2 is equal to σ_3 or σ_1. This way, the system allows for rotational symmetry, it is of constant use to admit that the principal stress $\sigma_1 = \sigma_r$ has the same value as the intermediate stress $\sigma_2 = \sigma_0$ perpendicular to every meridian plane.

In two-dimensional studies, there exist only two principal stresses σ_1 and σ_2.

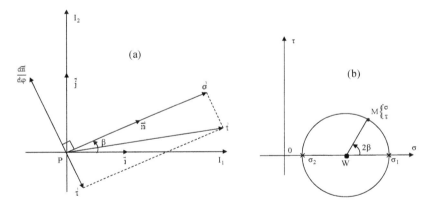

Figure 1.4. a) Principal coordinate system and b) coordinate system (σ, τ)

Let the unit vector $\vec{n}$ of the plane I_1PI_2 describe the circle of center P and let's locate it using its polar angle β with respect to the principal direction PI_1. The direction PI_2 is the other principal direction. The coordinates of $\vec{n}$ are:

$$\cos\beta \quad \text{and} \quad \sin\beta$$

Let i_1 and i_2 be the unit vectors in the principal directions.

Then:

$$\vec{n} = \vec{i}\cos\beta + \vec{j}\sin\beta$$

Let τ be the algebraic measure of the shear stress on the unit vector $-\dfrac{d\vec{n}}{d\beta}$ derived from $\vec{n}$ by the rotation $-\dfrac{\pi}{2}$.

The stress $\vec{t}$ exerted on the plane whose normal is $\vec{n}$ is written as:

$$\vec{t} = \sigma\vec{n} - \tau\frac{d\vec{n}}{d\beta}$$

or:

$$\vec{t} = \sigma\cos\beta i + \sigma\sin i$$

$$\vec{t} = \sigma\left(\cos\beta\vec{i} + \sin\beta\vec{j}\right) + \tau\left(\sin\beta\vec{i} - \cos\beta\vec{j}\right)$$

Comparing the coefficients of i and $\vec{j}$ in these two equations gives:

$$\sigma_1 \cos\beta = \sigma\cos\beta + \tau\sin\beta$$

$$\sigma_2 \sin\beta = \sigma\sin\beta - \tau\cos\beta$$

From where we obtain:

$$\sigma = \sigma_1\cos^2\beta + \sigma_2\sin^2\beta = \frac{\sigma_1 + \sigma_2}{2} + \frac{\sigma_1 - \sigma_2}{2}\cos 2\beta$$

$$\tau = \left(\sigma_1 - \sigma_2\right)\sin\beta\,\cos\beta = \frac{\sigma_1 - \sigma_2}{2}\,\sin 2\beta$$

In the σ, τ coordinate system, the point M image of the point P of the solid describes a circle centered at $\dfrac{\sigma_1 + \sigma_2}{2}$, of radius $\dfrac{\sigma_1 - \sigma_2}{2}$ and with center W. The polar angle of WM is 2β.

Note that the shear is highest for the directions of polar angles $\pm\dfrac{\pi}{4}$ with respect to PI_1. It is the bisecting direction with respect to the principal directions i_1 and i_2.

Thus, every image point P of coordinates τ and σ is located on a circle of radius $(\sigma_1 - \sigma_2)/2$ and whose center is on the abscissa $(\sigma_1 + \sigma_2)/2$. Every point of the circle is the *image* of a direction at a given point M of the solid. This direction is perpendicular to the area on which the stresses σ and τ are exerted.

Indeed, in plane coordinates and at the point M of the solid, we can define the direction of the normal to every plane passing through M with the help of the oriented angle β, which we have thus defined as:

$$\beta = \left(\vec{n}, \overrightarrow{o\sigma_1,} \right)$$

The normal and tangential components of the stress at M of this plane are respectively σ and τ.

If, on the Mohr's representation we conserved the angles' orientations, we just saw that we would have $-\tau$ on the ordinate axis. But practice wants that we use $+\tau$, which amounts to effecting a symmetry with respect to the $o\sigma$ axis and reversing the orientation of β. On this traditional representation, the oriented angles are therefore:

$$\beta' = -\beta = \left(\overrightarrow{o\sigma_1}, \vec{n} \right)$$

This convention is advantageous if we're looking to find the direction of a unit vector $\vec{n}$ with respect to the direction of the major principal stress σ_1.

Recall that τ, in the physical solid, is measured positively on the axis deduced from the entering normal by rotation of $+\pi/2$.

Note that the angles β and β' would not be modified if the *tensile* stresses were counted positively and if we considered the *outgoing* normal. Indeed, the two vectors $\vec{n}$ and $\overrightarrow{o\sigma_1}$ would be replaced by vectors of the same direction, but of opposite orientation.

NOTE.–

We say that a sample is all the more consolidated that the compression stress to which it was subjected was higher and the porosity was low. It is therefore compact but this state can equally be obtained by compression (for example with the help of tapping on the testing burette).

On the contrary, a less consolidated product is said to be mobile or loose and its porosity is high.

The *critical porosity* does not vary during a shear.

1.2.7. *Three-dimensional mechanics with symmetry plane*

Now let α, β, and γ be the direction cosines of the entering normal to a surface at the point M. The components t_x, t_y, and t_z of the stress vector applied to this area are such that:

$$\begin{bmatrix} t_x \\ t_y \\ t_z \end{bmatrix} = \begin{bmatrix} \sigma_x\ \tau_{yx}\ \tau_{zx} \\ \tau_{xy}\ \sigma_y\ \tau_{zy} \\ \tau_{xz}\ \tau_{yz}\ \sigma_z \end{bmatrix} . \begin{bmatrix} \alpha \\ \beta \\ \gamma \end{bmatrix}$$

We show that the stress matrix is symmetrical in three dimensions just like in two dimensions (see Laroze [LAR 74]).

The configuration of the containers (hopper or silo) differs depending on whether the horizontal section of the convergent (and the cylinder which surmounts it) is rectangular or circular.

If the section is rectangular and the largest side is at least equal to three times the smallest, it suffices to describe the stresses in a vertical parallel plane to the smallest side (in plane coordinates).

If the section is circular and the container has an axis of symmetry, it suffices to describe the stresses in a meridian plane.

Assume that:

– for a horizontal rectangular section (plane rectangular coordinates):

- the x-axis is horizontal and parallel to the smallest side, toward the right;

- the y-axis is vertical toward the top.

– for a horizontal circular section (cylindrical rotational coordinates):

- the r-axis is horizontal and its origin is on the considered solid's axis;

- the z-axis is vertical toward the top.

The "θ" axis completes the right trihedron.

In both cases, it is sufficient to describe the constraints in a plane of symmetry (meridian plane or parallel plane to the smallest side).

1.2.8. Displacement laws [JEN 60]

For the D.S. to *begin* to lose its shape, a circle of Mohr needs to become tangential to the static rupture curve. For a movable solid, without cohesion and unconsolidated, the rupture point reduces to $\tau = \sigma = 0$ (see Figure 1.5).

The direction of the displacement vectors $\vec{\varepsilon}$ and $\vec{\gamma}$ correspond respectively to the directions $\vec{\sigma}$ and $\vec{\tau}$. The global vector $\vec{e} = \vec{\varepsilon} + \vec{\gamma}$ is perpendicular to the static rupture curve. Let φ_v be the angle which the direction $\vec{\gamma}$ (which is also that of $\vec{\tau}$) makes with $\vec{e}$. In general, φ_v is positive for $\gamma > 0$ and $\varepsilon < 0$. The fact that the expansion ε is negative in the direction of compressions indicates that the solid is expanding. Note that φ_v is equal to the angle of internal friction $\varnothing$.

In reality, Nemat–Nasser [NEM 80] introduced what he called the dilatancy angle v, which adds up to the angle φ_v for displacements. A positive value of the angle v increases expansion and a negative value of the angle v corresponds to a contraction.

After an expansion, the mass density of the solid reduces and its rupture curve draws closer to the $o\sigma$ axis. As a result, there is a jump to a new rupture curve.

If we consider the rupture curve, the angle φ_v reduces as we draw closer to the terminal point E. Then, the jump is almost insignificant and a continuous plastic flow appears at E without changing the porosity.

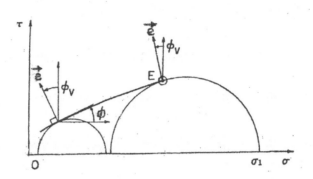

Figure 1.5. *Direction of displacements*

1.3. Flow of divided solids

1.3.1. *Preconditioning and critical condition*

Preconditioning refers to the set of operations that lead to a definite consolidation of the sample before testing begins.

In Jenike's shearing cell, we fill the layers successively which we level out with a number of rotations of the lid. We then exert the vertical stress $\sigma_v = F_v/A$. In this way, we obtain the critical porosity ε_c corresponding to the preconsolidation stress σ_v of the sample, which will then be subjected to shear deformation under stresses inferior to σ_v.

Roscoe *et al.* [ROS 58] (Figure 33 of his book) describe a trial in which a series of samples have undergone different consolidations, such that, even if they are subsequently subjected to the same confining pressure σ_c, their porosities ε_0 are different.

Conversely, if all these samples are subjected to shear tests at different confining pressures inferior to the consolidation pressure σ_c, we see that the porosity of the samples tends toward a common value ε_c which is the *critical porosity corresponding to the prior consolidation pressure*.

The form of the equation $\varepsilon_c = \varepsilon_c (\sigma_c)$ obviously depends on the product considered but unfortunately too on the type of equipment used, as the definition of the confining stress varies from one equipment to the other.

Particularly, the state of a D.S. depends on the direction of the principal stresses to which it has been subjected.

The confining stress on the walls of a silo is approximately perpendicular to the shear stress in the mass of the product. However, in the shearing cells, the consolidation stress (vertical) is also perpendicular to the shear stress (horizontal). It can be shown that this coincidence makes the cell results valid for the calculation on the silos.

1.3.2. Experimental results if $\sigma_c \neq \sigma_c^*$

Assume that the confining stress σ_c used during testing does not have the critical value σ_c^* corresponding to the critical porosity (see Figure 1.6).

If $\sigma_c < \sigma_c^*$, the sample is said to be overconsolidated relative to the considered test.

If $\sigma_c > \sigma_c^*$, the sample is said to be underconsolidated relative to the considered test.

Structural considerations allow one to predict what will be the behavior of a D.S. depending on its state.

If the sample is overconsolidated, the particles will be more or less entangled and nested one inside the other. For a shear deformation to be possible, it would have to be followed by a lateral displacement (expansion), in such a way that the particles have the space to leapfrog one another. The expansion will be all the more important as the initial tangle will be brought out more. Moreover, this expansion reduces the density of mutual contacts between particles and consequently, the shear stress will reduce in the course of the test.

If, on the contrary, the sample is underconsolidated, its structure becomes very aerated and can fall apart, thus contracting as soon as the confinement stress is applied. The number of mutual interactions between particles increases and the shear stress increases in the course of the test.

Naturally, if $\sigma_c = \sigma_c^*$, the sample is said to be in the critical state and the shear stress remains constant.

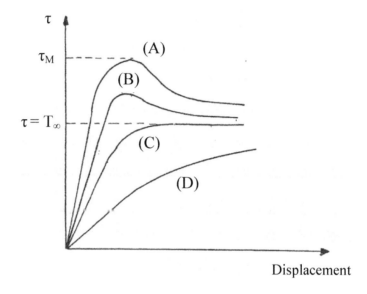

Figure 1.6. *Variation of shear stress with respect to displacement*

(A) and (B): overconsolidation;

(C): critical condition;

(D): underconsolidation;

τ_c = constant for each of the four cases;

[$\tau = \tau (\sigma_c)$ is defined by the value of the confining stress σ_c].

When the sample is overconsolidated, a slight complication arises. It always starts behaving as if it was underconsolidated with initial increase in the shear stress, which is greater than the critical value. After crossing a maximum value τ_M, its behavior is that which we have described.

From the previous structural interpretation, it follows that:

– if a sample is underconsolidated, it contracts and its volume reduces just as its porosity;

– if a sample is overconsolidated, it expands and its volume increases just as its porosity.

For a large enough shear displacement, the porosity of the D.S. always tends towards a critical value ε^* which is a function of the prior consolidation stress.

Recall that the prior consolidation stress σ^* is different from the stress σ under which the displacement is made:

– if $\sigma^* > \sigma$, there is overconsolidation;

– if $\sigma^* < \sigma$, there is underconsolidation.

Roscoe *et al.* [ROS 58] represented in a trirectangular trihedron the variation of the porosity ε (or rather of the ratio $e = \dfrac{\varepsilon}{1-\varepsilon}$) in terms of the stresses σ and τ and they observed that the limiting value of ε (the critical value ε^*) moved along a curve not in a plane in this trirectangular trihedron. The same observation will be made on the Figure 5 of York [YOR 80].

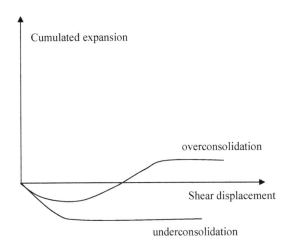

Figure 1.7. *Variation of the volumetric expansion of a D.S. according to its consolidation state*

Practically, the preconditioning that is the prior consolidation is constant and we vary σ. We notice moreover with Jenike that every product which flows from a hopper is *overconsolidated* by residing in the hopper. We will then come back to the behavior of a consolidated product in a shear cell.

As soon as the displacement of the shear cell's ring starts, the shear resistance increases almost instantaneously up to a maximum value τ_M and this maximum value is attained while the displacement is very limited. In fact, it's about a course that is elastic in nature. The product, during this phase, "accepts" the stresses imposed on them and contract slightly. Subsequently there is "rupture", that is unblocking of particles, disentanglement and expansion of the product. In the structure thus "loosened", the shear stress τ reduces while the displacement continues. This course is plastic in nature.

The maximum level τ_M of τ increases slowly with the confinement stress σ_c. When the prior consolidation increases, we know that τ_∞ doesn't vary, but on the other hand, the maximum value τ_M increases slightly.

NOTE.–

The curves (A) and (B) of Figure 1.6 are such that:

$$\tau_M^{(A)} > \tau_M^{(B)} \quad \text{and} \quad \sigma^{*(A)} > \sigma^{*(B)}$$

τ_∞ is the shear stress limit of an overconsolidated product when σ_c draws closer to σ_c^*.

NOTE.–

Nemat-Nasser [NEM 80], in his theoretical study of dilation, rationally introduces the notion of dilatancy angle. However, the scope of this study is limited to products without cohesion.

NOTE.–

Nemat-Nasser [NEM 80] confirms in his Figure 4 that the dilation follows and analog variation to that of τ in terms of the shear displacement (see Figure 1.6).

Similarly, York [YOR 80], in his Figure 4, confirms the curves of Figure 1.6.

1.3.3. *Trial procedure according to Jenike*

This author admits, as we wanted, that every product which flows from a hopper is overconsolidated such that all the trials corresponding to a given consolidation are carried out at confinement pressures σ_c *inferior* to that which led to the consolidation during preconditioning, that is σ_c^*.

The product being overconsolidated, we notice that the horizontal resistant force F_H reaches a maximum during the translational movement of the ring.

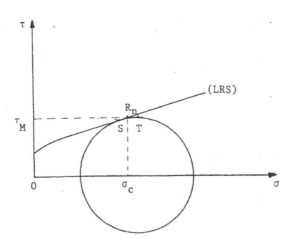

Figure 1.8. *Image of a trial on a Mohr's circle*

We begin trials with $\sigma_c^{(1)}$ being the least expected value. We continue by increasing the value of σ_c and, for the last trial, $\sigma_c^{(n)} = \frac{3}{4}\sigma_c^*$.

Jenike explains that the couple of values $\tau_M^{(i)}\sigma_c^{(i)}$ corresponding to the i-th trial is located on Mohr's circle of stresses between the highest point of the circle T and the contact point S of the circle and rupture point. We then make a slight error admitting that the image R_i ($\tau_M^{(i)}$, $\sigma_c^{(i)}$) is located on the static rupture curve (LRS).

The series of trials carried out provides n points R_i, which enable us to plot the LRS corresponding to a specific prior consolidation state (preconditioning), which defines the limiting point E.

1.3.4. Rational evolution of consolidation

We define, according to Jenike, a consolidation by the value of the major principal stress σ_1 supported by the sample during the preconditioning.

We derive this stress by extrapolating the LRS up to the extreme point E corresponding to the preconditioning consolidation and drawing Mohr's circle tangent to the LRS at the point E. Let's specify the nature of the extreme point E (see Figure 1.10).

For every value of σ_1 inferior to the consolidation's σ_1, when the Mohr's circle draws closer to the LRS, we witness a jump in the value of the shear stress. But, as we draw closer to E, the magnitude of the jump reduces and, when σ_1 meets the consolidation point, there is no longer a jump but rather a continuous plastic flow.

The intersection of this circle with the 0σ axis gives graphically, to the right of the circle's center, the value of σ_1.

1.3.5. Hysteresis

If, for example, in an oedometer, we carry out a simple compression trial on a sample and we then cancel the compressive force, the product doesn't regain its entire initial volume.

We notice the existence of a residual contraction (hysteresis) reflecting the irreversible nature of the sliding and mutual reorientation of particles during compression. The new disposition thus obtained is more stable.

In plane elasticity, it follows that a rotation of $\pi/2$ of the principal stresses is usually followed by a contraction. Indeed, when the major principal stress reduces and becomes minor, the regain in volume in this direction is limited.

On the contrary, when the minor principal stress becomes major, the contraction is maximal.

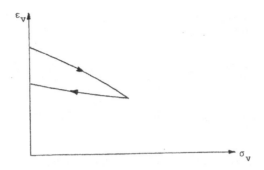

Figure 1.9. *Simple compression cycle*

1.3.6. *Interpretation of results: non-confined yield stress*

The quasi-static rupture region reacts to a given preconditioning, that is a given preconditioning defined by the major principal stress σ_1.

We know that a Mohr's circle cuts the 0σ axis at σ_1 (major principal stress) when it is tangent to the LRS at the extreme point E.

But another Mohr's circle is drawn which is tangent at the same time:

– to the LRS;

– to the 0τ axis at the origin.

Its second intersection with the 0σ axis gives *the non-confined yield stress*, f_c (see Figure 1.10).

The larger part of the LRS can be compared to a straight line. This straight line makes an angle ϕ with the 0σ axis. This angle is the internal friction angle for a solid initially at rest at preconditioned at E.

The LRS corresponds to the rupture of the static equilibrium that exists before motion starts. The intersection of the straight line equivalent to the 0τ axis is the "cohesion" that determines the amount of stress required to begin a shear movement and separate the particles all the more highly entangled as σ_1 was higher.

The area contained in the concave part of the LRS is the rigidity domain (also called as elastic area) where only one system of limited displacements in space and time can correspond to a given system of stresses. More specifically, in our own opinion, an elastic solid is slightly deformed and in a reversible manner.

When Mohr's circle is tangent to the LRS at the point E, the product's behavior is no longer rigid but plastic in nature and displacements are made under constant stresses (Tresca [TRE 68] made accurate tests on plastic deformation.)

Von Mises [VON 13], as well as Haar and Von Karman [HAA 09], carried out a theoretical study of plastic deformation.

When we vary the initial consolidation stress σ_1, it is tantamount to subjecting the previous representation to a homothety of center 0 and, for example, if σ_1 increases:

– the rigidity area increases;

– the point E moves along a straight line passing through the origin and that is the curve of dynamic rupture;

– cohesion increases proportionately with σ_1;

– the nonconfined stress also increases proportionately with σ_1;

– the internal friction angle ϕ practically remains constant.

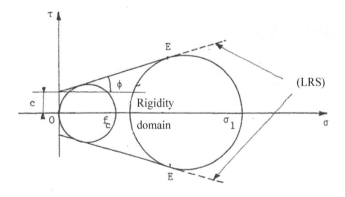

Figure 1.10. *Static rupture region (LRS)*

That is, however, about an approximation and, in reality, f_c increases slightly less proportionately with σ_1. Not too close to the origin, we can compare the curve $f_c (\sigma_1)$ to a straight line and say that:

$$f_c = f_{co} + k\sigma_1$$

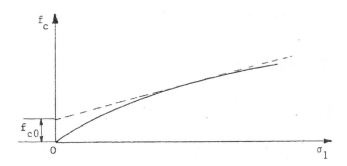

Figure 1.11. *Variation of $f_c (\sigma_1)$*

1.3.7. *Flowability and flow coefficient*

A D.S. which is "flowing" a lot can easily leave the rigid state and become deformed. In other words, a flowing product possesses a limited rigidity area. But, a good measure of this area is the flowability C:

$$C = \frac{1}{f_c}$$

The flowability is, in terms of units, the reciprocal of pressure and is measured in Pa^{-1} or, which is actually the same thing, in $m.s^2.kg^{-1}$. The flowability of a D.S. is different from a Newtonian liquid's "fluidity", which is simply the reciprocal of viscosity and is measured in $m.s.kg^{-1}$ (reciprocal of Pa.s).

Considering the expression for f_c, which will be proven later, we can also say that (see section 1.4.2):

$$C = \frac{1}{2c \tan\left(\dfrac{\pi}{4} + \dfrac{\phi}{2}\right)}$$

From this expression, we see that a product is particularly more flowing as its internal friction angle ϕ and above all its cohesion c tend toward zero.

Jenike assumed, in his first approximation, that f_c was simply directly proportional to σ_1 and introduced a flow coefficient which he denoted as ff_c and which we will call k_E:

$$k_E = \frac{\sigma_1}{f_c}$$

It is also written as:

$$C = \frac{1}{f_c} = \frac{k_E}{\sigma_1}$$

Recall that σ_1 is the major principal stress at a given point. But, this stress varies from one point to another of the solid such that the flowability of a product is not an intrinsic property of the product and varies with the chosen point according to the local value of σ_1. On the contrary, the flow coefficient k_E varies a lot less with σ_1.

The preconditioning discussed above is, for us, essentially that of a product, which has been in a hopper or a silo. The more a product is preconsolidated (high values of σ_1) the less flowing it is (C is low).

The following table gives some orders of the magnitude of k_E.

Flow coefficient k_E	Flowability class
$k_E < 2$	Always nonfluent
$2 < k_E < 4$	Faintly fluent
$4 < k_E < 10$	Very fluent
$k_E > 10$	Always perfectly (freely) fluent

Table 1.1. *Flow coefficient*

Associating a "fluent" property to the coefficient k_E obviously means that we always refer to the usual range of σ_1 for divided solids (2,000–50,000 Pa).

The flow coefficient k_E (or $f\!f_c$) especially has a qualitative interest and accurate calculations require a complete knowledge of the function f_c (σ_1). The following figure gives some examples.

This figure enables us to predict serious flow difficulties for fibers (glass or asbestos).

1) Very fluent or freely flowing: $k_E \geq 7$

Cereal grains, plastic pellets, sugar, salt (dry), packed charcoal, graphite.

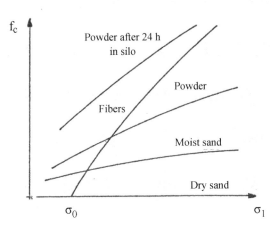

Figure 1.12. *Limiting unconfined stress f_c (σ_1)*

2) Relatively fluent: $4 < k_E < 7$

Unsieved coal, unriddled bauxite, limestone, fertilizer, phosphate, angular gravel, moderately consolidated (packed) powders such as plant flour and minerals, cocoa, coloring agents (soluble in water), and pigments (insoluble). Common pigments include titanium oxide, iron oxide, carbon black, zinc oxide, etc.

3) Very slightly or non-fluent: $k_E < 4$

Raw coal ore comprising of grained and fine. Coke, dairy, slag, clinker, raw coal ashes, plant coat cracks, sawdust, wood chips, fibers, flakes, plastic or rubber waste, wet crystals (85–90% saturation).

It is *important* to note that the classes thus defined only correspond to a first approximation.

NOTE.–

Schulze [SCH 98] studied what he called the flow characteristic.

1.3.8. Rigidity, plasticity, flowability, and mobility

The rigidity domain of a D.S. is defined in the following manner in a Mohr's coordinate system:

– the extreme point E (σ, τ) and the Mohr's circle associated to it set the conditions for initial consolidation (in particular the major principal stress σ_1);

– the cohesion c;

– the internal friction angle ϕ.

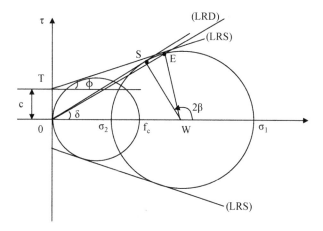

Figure 1.13. *Rupture areas*

The rigidity domain will be, for us, defined by the two inequalities:

$$\sigma > 0 \quad \text{and} \quad |\tau| < c + \sigma \, tg\phi$$

When the second inequality becomes an equality, the equilibrium is said to be limited and the product then suffers shear failure. A Mohr's circle tangent to the 0τ axis and to TE recuts the 0σ axis in a point of abscissa f_c, which is the limiting unconfined stress.

Now let us look at the behavior of a D.S. which is no longer in a state of failure, but in a steady flowing state. A product in steady flowing state is such that the shear slips acting within it occur following a direction whose $\vec{n}$ is such that the oriented angle β of $\vec{W\sigma}$ with $\vec{n}$ is equal to $\pm\dfrac{1}{2}\left[\dfrac{\pi}{2}+\phi\right]$. If OS is the tangent from O to the Mohr's circle corresponding to the flowing product's consolidation state, then:

$$\vec{W\sigma_1},\ \vec{WS}=2\beta$$

Experience shows that, regardless of the value of σ_1, the points E are located on a half-line thus defined:

$$|\tau|=\sigma\mathrm{tg}\delta \quad \text{with } \sigma>0$$

δ is the dynamic friction angle (also called "efficient"). We can now define flowability and mobility.

Due to the stability of the angle δ, f_c, and c are directly proportional to σ_1. But, cohesion helps us to determine whether a product subjected to a pure shear stress will break down or not and begin to flow. We know that the product's flowability is given by $C = \dfrac{1}{f_c} = \dfrac{k_E}{\sigma_1}$.

Once the product has been broken down, it starts moving under the influence of an external field of force (usually gravity). This movement is particularly simple as the dynamic friction angle δ is weak and we define mobility by (see Figure 1.13):

$$m=\frac{\sigma_2}{\sigma_1}=\frac{OW(1-\sin\delta)}{OW(1+\sin\delta)}=\frac{1-\sin\delta}{1+\sin\delta}$$

For most products, δ is found between $30°$ and $60°$ such that the mobility varies between 0.3 and 0.03. Mobility represents the ratio σ_2/σ_1 of the minor and major principal stresses corresponding to the consolidation. However, the parameter m is especially an arithmetic trick such that:

for a rigid solid $\delta=\pi/2$ m = 0

for a liquid $\delta=0$ m = 1

1.3.9. Energy dissipation

When a D.S. loses its shape by flowing away, the shear stresses required for deformation are all the more intense as the angle δ is large. Thus, a high value of the angle δ has two implications:

– the energy dissipation in the flowing product is amplified;

– we see from Mohr's representation that the cohesion as well as f_c increase.

In other words, the less flowing product will have the tendency:

– to move as a unit without losing its shape;

– and, if it loses its shape, to dissipate energy.

1.3.10. Analytical expression of rupture regions

The following expression, which is now classical, was suggested and very utilized by Coulomb [COU 73]:

$$\tau = c + \sigma tg\phi$$

c represents the cohesion and ϕ the internal friction angle.

For negative normal stresses (traction), some make use of a purely empirical formula. For example:

$$\left[\frac{\tau}{c}\right]^n = \frac{\sigma + \sigma_o}{\sigma_o}$$

Near the origin, we can compare the LRS to a parabola tangent to Mohr's circle of diameter f_c and passing through the origin, the slope of the tangent being tg ϕ. As a result:

$$\sigma_o = f_c \left[\frac{1 - \sin\phi}{4 \sin\phi}\right]^2$$

From Jenike's Figure 6 [JEN 59], we see that Coulomb's criterion is written as:

$$ACsin\phi = CE = \frac{\sigma_1 - \sigma_2}{2}$$

ϕ: internal friction angle.

But, from section 1.2.4,

$$CE = \frac{\sigma_1 - \sigma_2}{2} = \sqrt{\frac{1}{4}\left(\sigma_x - \sigma_y\right)^2 + \tau_{xy}}$$

and

$$\sigma_x + \sigma_y = \sigma_1 + \sigma_2$$

Finally, we deduce from Jenike's Figure 6 [JEN 59]

$$AC = \frac{\sigma_x + \sigma_y}{2} + c\,ctg\phi = \frac{\left[\frac{1}{4}\left(\sigma_x - \sigma_y\right)^2 + \tau_{xy}^2\right]^{1/2}}{sin\phi}$$

which the expression of Coulomb's criterion for a coordinate system xoy.

Now let us look at the dynamic rupture region (LRD). Let us assume that we have established a family of LRS for different conditions and that the extreme points E are known.

The location of these extreme points is the LRD. It is a straight line passing through the origin. Indeed, a solid undergoing deformation would not experience cohesion as cohesion defines the transition from mobility to deformation. Experience shows that the LRD consists of two half-lines coming from the origin and making an angle δ, which is the dynamic internal friction angle, with the 0σ axis. Jenike states: "efficiency" (understated: for flow).

The LRD is expressed with respect to the principal axes $O\sigma_1$ and $O\sigma_2$ Jenike [JEN 62].

$$sin\delta = \frac{\sigma_1 - \sigma_2}{\sigma_1 + \sigma_2}$$

Provided $\sigma = \dfrac{\sigma_1 + \sigma_2}{2}$, this equation can be written as:

$$(\sigma_1 - \sigma)^2 + (\sigma_2 - \sigma)^2 = 2\sigma^2 \sin^2 \delta$$

The generalization of this equation in three dimensions is immediate.

$$(\sigma_1 - \sigma)^2 + (\sigma_2 - \sigma)^2 + (\sigma_3 - \sigma)^2 = 2\sigma^2 \sin^2 \delta \quad \text{with} \quad \sigma = \dfrac{\sigma_1 + \sigma_2 + \sigma_3}{3}$$

The shear tests give, for a given product:

– the family of different LRS of which each corresponds to a specific principal stress σ_1 (given by Mohr's circle tangent to the LRS at the extreme point E).

– the overall LRD.

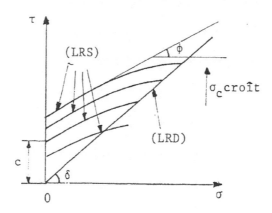

Figure 1.14. *Network of characteristic curves of a product σ_c is the consolidation stress*

1.3.11. Influencing parameters on the friction angles

The analytic equations of the rupture regions are:

LRS $\quad |\tau| = c + \sigma \, \mathrm{tg}\phi$

LRD $\quad |\tau| = \sigma \, \mathrm{tg}\delta$

By definition, the LRD is independent of the prior consolidation.

The LRS represents the rigidity domain corresponding to a prior consolidation and, as we already know, this domain increases with the consolidation.

Now let us look at the influence of the two parameters:

1) The particle size and granulometry (size distribution):

If the granulometry is spreading, the mutual interlocking of particles is high as the fine particles will slip between the grained ones, which reduces the porosity and increase the cohesion as well as the unconfined stress f_c. Finally, the rigidity domain is increased.

The dynamic friction angle δ also increases with the spreading of granulometry, that is the proportion of fine particles.

2) Sphericity and angularity of particles:

If the shape of the particles moves away from that of a sphere, there are two contradictory consequences:

– the product's porosity increases, which reduces the mutual interlocking and entanglement of the particles;

– on the other hand, a less spherical shape increases the entanglement of particles.

Thus, in a fiber bundle, the porosity is very high (can be higher than 98%) but the elasticity domain is increased (high cohesion).

A clump of plates also possesses a high porosity (80%) and moreover, a reduced rigidity domain. It's the case of ground talc and mica.

Spheroidal angular or rough particles will sometimes have a low cohesion but a high dynamic friction angle.

NOTE.–

Thornton [THO 93] studied the relationship between the surface energy of particles and the modulus of elasticity of the resulting solid by putting the particles in contact (interest in sintering).

1.4. Current identities

1.4.1. *Relationship between the principal stresses and the unconfined stress*

The static rupture line (LRS) (in terms of locus), which defines the limit state, consists of two half-straight-lines in Mohr's representation.

The position of these half-straight-lines depends on the σ_1 value of the major principal stress in the product during preconditioning (confinement stress). These value σ_1 constitutes the upper limit above which the rupture region is no longer applicable (extreme point E on the representation).

These rupture lines are therefore only concerned with states subjected to a major principal stress less than σ_1. For all these states, the product's behavior is therefore of overconsolidated type. Indeed, a study of the stored product consolidated in a hopper's arching only makes use of such stresses.

Let's see how this can be seen on Mohr's circle.

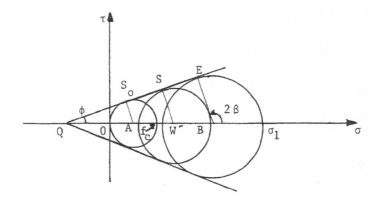

Figure 1.15. *Rupture region for a solid initially at rest*

φ is the internal friction angle in the traditional sense of the word.

The major principal stress σ_1 defines the rupture region and, for this maximum value of the stress, the angle β defines the direction of the normal to the product's internal sliding plane with respect to σ_1. This angle is the same for all the limit circles of Mohr irrespective of the value of σ_1 (homothety to Q as its center):

$$2\beta = (W\sigma_1, WS) = \text{const.}$$

We can write:

$$\sin\phi = \frac{WS}{QW} = \frac{AB/2}{QO + \dfrac{0A + 0B}{2}} = \frac{(\sigma_1 - \sigma_2)/2}{QO + \dfrac{\sigma_1 + \sigma_2}{2}} \qquad [1.3]$$

For the particular circle such that $\sigma_2 = 0$, we can say that:

$$\sigma_1 = f_c$$

Hence:

$$\sin\phi = \frac{f_c/2}{QO + f_c/2} \qquad \text{hence}: 2QO = \frac{f_c(1 - \sin\phi)}{\sin\phi} \qquad [1.4]$$

Eliminating QO from [1.3] and [1.4]:

$$\sigma_1 = f_c + \frac{1 + \sin\phi}{1 - \sin\phi}\sigma_2$$

When an arch is formed, the major principal stress σ_1 is parallel to its surface and the minor principal stress σ_2 which is perpendicular to it, is zero on its bottom surface which is free (unconfined). That is why $\sigma_1 = f_c$ is referred to as the unconfined stress.

1.4.2. Other current identities

In this section, we shall call δ the angle, which Oσ makes with the tangent OS to Mohr's circle. This angle is not the dynamic friction angle, which involves the straight line OE and which we have defined in section 1.3.8. However, some authors find it convenient to mix the two angles up.

The only difference between the Figures 1.13 and 1.16 is that, in Figure 1.16, we have drawn a line OP parallel to TS and passing through O and that we have deleted the circle of diameter Of_c.

1) The average principal stress is given by: $\bar{\sigma} = OQ$

The radius of Mohr's circle is: $r = QH = \bar{\sigma}\sin\delta$

The principal stresses are:

$$\sigma_1 = \bar{\sigma} + r = \bar{\sigma}(1+\sin\delta)$$

$$\sigma_2 = \bar{\sigma} - r = \bar{\sigma}(1-\sin\delta)$$

$$\frac{\sigma_1}{\sigma_2} = \frac{1+\sin\delta}{1-\sin\delta} \qquad\qquad [1.5]$$

2) The cohesion being c = TO and the straight lines TE and OP being parallel:

$$EP = c\cos\phi = QE - QP = r - \bar{\sigma}\cos\phi = \bar{\sigma}(\sin\delta - \sin\phi)$$

$$c = \sigma_1\frac{(\sin\delta - \sin\phi)}{(1+\sin\delta)\cos\phi} \qquad\qquad [1.6]$$

3) We can write:

From $[1.5]$ and $[1.6]$: $c\cos\phi = \sigma_2\left(\dfrac{\sin\delta - \cos\phi}{1-\sin\delta}\right)$ $\qquad [1.7]$

From $[1.5]$: $\sin\delta = \dfrac{\dfrac{\sigma_1}{\sigma_2}-1}{\dfrac{\sigma_1}{\sigma_2}+1}$ $\qquad\qquad [1.8]$

Eliminating sin δ from [1.7] and [1.8]:

$$\frac{\sigma_1}{\sigma_2} = \frac{2c\,\cos\phi}{\sigma_2(1-\sin\phi)} + \frac{1+\sin\phi}{1-\sin\phi} \qquad [1.9]$$

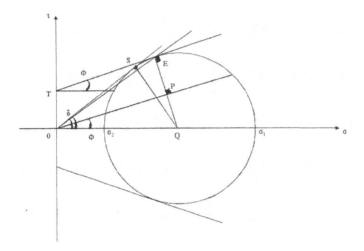

Figure 1.16. *Current identities*

4) The angle STO is (Figure 1.17):

$$\mathrm{STO} = \frac{\pi}{2} + \phi$$

From the triangle TOW, we obtain:

$$\mathrm{OTW} = \frac{\mathrm{OTS}}{2} = \frac{\pi}{4} + \frac{\phi}{2}$$

And:

$$\mathrm{OW} = \frac{f_c}{2} = \mathrm{TO}\ \mathrm{tg}(\mathrm{OTW}) = c\ \mathrm{tg}\left(\frac{\pi}{4} + \frac{\phi}{2}\right)$$

$$f_c = 2c\ \mathrm{tg}\left(\frac{\pi}{4} + \frac{\phi}{2}\right) \qquad [1.10]$$

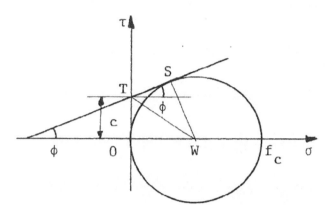

Figure 1.17. *Mohr's circle for the unconfined stress*

1.4.3. *Friction angle on a surface [OOM 85]*

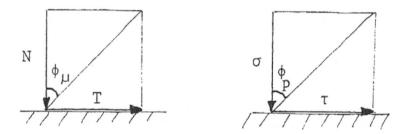

Figure 1.18. *Force and stresses on a wall*

The friction angle of a solid on a plane surface can be defined in two ways:

– if the solid is compact (a ball for example), the friction angle ϕ_p is:

$$\tan\phi_p = \vec{T} / \vec{N} = \tan\phi_\mu$$

$\vec{T}$ and $\vec{N}$ are the tangent and normal forces exerted on the surface by the solid when it starts moving. The friction angle depends only on the nature of the surfaces. It is denoted as ϕ_μ.

– if the solid is a D.S., we apply the previous forces to the unit surface area by taking the normal stress σ and shear stress τ into consideration:

$$\sigma = N \, / \, S \quad \text{and} \quad \tau = T \, / \, S$$

$$\tan\phi_p = \tau \, / \, \sigma$$

A slide is especially *easy* as the friction angle ϕ_p is less.

We will henceforth limit ourselves to divided solids.

1) Effect of the shape of particles (angularity) and their roughness:

When a D.S. slides along a surface, its particles slide or roll along the surface with it. The amount of particles that roll depends on the sphericity of these particles, that is the resemblance of their shape with that of a sphere.

Less spherical particles will have a greater tendency to slide rather than roll and this leads to a second property of particles, which is their angularity and their roughness, that is their tendency to possess sharp edges or sharp corners on their surface. The sliding of angular particles can easily be accompanied by a jamming like engineers call it, and the sharp corners and sharp edges have a tendency to cut and pierce through the surface. In this case, either the particles are deformed and even broken or their surface, if it is softer than the particles, it gets scratched. The coefficient of friction of angular gravel can reach a value of 0.45.

In short, the sliding of a D.S. will be eased by a large sphericity and a low amount of roughness of its particles. The friction angle ϕ_p will also be low.

2) Effect of the size of particles d_p:

The size of particles has an effect all the more important as their sphericity is high. In fact, the larger particles slide more easily because their center of inertia is further away than their rolling surface. Thus, the torque of the lever arm is greater. We can also say that the friction on the wall increases especially as d_p is less than the natural roughness of the wall as we will see below.

According to the DIN 1055 standard of 1963, we refer to:

– granular divided solids as those solids whose particles have sizes greater than or equal to 200 µm. These are the granular solids (or graules).

– powdery divided solids as those solids whose particles have sizes less than or equal to 60 μm. These are the fine solids.

However, according to this standard, the ratio of friction angles for the granular solids to the powder solids is given by:

$$\frac{\phi_{pg}}{\phi_{pp}} = 0.75$$

3) Influence of the roughness of the wall:

For a dry surface and a "polished" surface (whose absolute roughness is much less than the size of particles):

$$8° < \phi_p < 13°$$

This value of ϕ_p draws closer to the value of ϕ_μ mentioned earlier.

As the roughness increases, the angle ϕ_p increases and when the absolute roughness becomes equal to the average diameter of the particles, the friction angle on the wall draws closer to the product's internal friction angle.

Thus, for most granular products, the following table gives magnitudes of the coefficient of friction on the wall of silos and hoppers.

If the renewal of the product is frequent, the walls are subjected to a polishing effect which can significantly reduce the friction coefficient. For powdery products such as cement or flour, the effect of pressure can hide that of the nature of the wall.

Nature of the wall	tg ϕ_p
Coated with polymer	0.17
Aluminium, stainless steel	0.25–0.35
Ordinary steel	0.35
Wooden panels, reinforced concrete	0.35–0.45
Walls with interior strapping	0.45–0.50
Corrugated sheet	0.50–0.60

Table 1.2. *Coefficient of friction on the wall*

Mac Atee *et al.* [MAC 91] give the precautions to be taken when coating the wall with a polymer.

4) Effect of pressure:

For a product without cohesion and less compressible, like dry sand, the friction angle on a wall is independent of the compression stress σ. On the other hand, for a compressible and cohesive product, like wheat flour, the friction angle decreases with σ as shown in the Figure 1.19.

The coefficient of friction of the cement can vary from 0.22 to 0.40 depending on the pressure. However, in a hopper or a silo, the pressure is maximum at the transition between the vertical part and the convergent part (of the order of thousands of Pascals) and it draws closer to zero at the outlet.

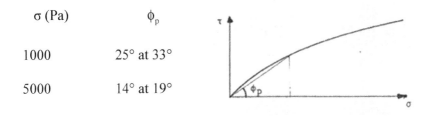

σ (Pa)	ϕ_p
1000	25° at 33°
5000	14° at 19°

Figure 1.19. *Friction on the wall*

5) Chemical nature of the product:

Diatomaceous earth (powdery in nature) has the same coefficient of friction on mild steel and stainless steel probably because a layer of soil is adsorbed on both surfaces, which makes their behavior similar. Similarly, attapulgite clay, which is powdery clay, has the same friction behavior on polished aluminium and on polished stainless steel. On the other hand, some salts have different coefficients, probably because they are adsorbed differently.

6) Sliding velocity:

Very low and of the order of 1 cm.s^{-1} on the vertical walls, the sliding velocity at the outlet opening can rise up to a value of 1 m.s^{-1}. However, this speed does not have a considerable effect on the coefficient of friction. On the contrary, the nearly constant value of the dynamic friction angle is 5°–10° less than the angle for which sliding begins from a static condition. The values frequently published in literature are those of the dynamic angle.

7) Some orders of magnitude (sexagesimal degrees):

Product	Concrete	Steel wall	Plastic
Limestone (0–56 mm)	33	29	17–19
Limestone (60–120 mm	37	29	15–16
Quicklime (50–200 mm)	35	28	15–17
Raw gypsum	36	35	11–14

Table 1.3. *Friction angle on the wall*

8) Measurement of the friction on the wall.

The angle ϕ_p can be measured with a shear cell by replacing the mobile base with a plate created with the wall material.

1.4.4. Sliding on a vertical wall: active and passive states

When the friction on the wall is fully involved, the following relationship exists between the normal stress σ and the shear stress τ exerted by the product on the wall.

$$|\tau| = \sigma \text{tg} \phi_p$$

ϕ_p is the friction angle on the wall.

Let us assume that Mohr's circle at a point M is known.

The abscissa of its center is defined by the average stress $\bar{\sigma}$.

$$\bar{\sigma} = \frac{\sigma_1 + \sigma_2}{2}$$

The radius of Mohr's circle is defined by:

$$r = \frac{\sigma_1 - \sigma_2}{2}$$

Let us bring down the perpendicular line WH from W on OA (see Figure 1.20):

$$WH = WA \sin \Delta = r \sin \Delta = OW \sin \phi_p = \overline{\sigma} \sin \phi_p$$

Therefore:

$$\sin \Delta = \frac{\sin \phi_p}{r / \overline{\sigma}}$$

The friction is fully involved if the product is in motion. Mohr's circle is therefore almost tangent to the LRD whose equation is:

$$|\tau| = \sigma \, \mathrm{tg} \delta$$

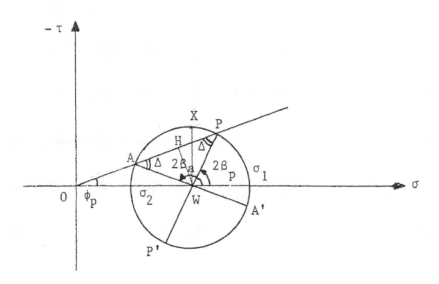

Figure 1.20. *Friction on the wall and Mohr's circle*

δ is the dynamic friction angle.

For such Mohr's circle:

$$\frac{r}{\overline{\sigma}} = \sin\delta \quad \text{and} \quad \sin\Delta = \frac{\sin\phi_p}{\sin\delta} \qquad\qquad [1.11]$$

At every point M of the moving solid, the properties of Mohr's circle are such that:

$$\sigma = \overline{\sigma} + r\,\cos 2\beta = \overline{\sigma}\left(1 + \sin\delta\cos 2\beta\right)$$

$$-\tau = r\sin 2\beta = \overline{\sigma}\,\sin\delta\sin 2\beta$$

And for the normal direction to the wall:

$$-\tau\,\cos\phi_p - \sigma\,\sin\phi_p = 0$$

$$\sin\delta\left(\sin 2\beta\cos\phi_p - \cos 2\beta\sin\phi_p\right) = \sin\phi_p$$

$$\sin\left(2\beta - \phi_p\right) = \frac{\sin\phi_p}{\sin\delta} = \sin\Delta > 0$$

We will choose the angle Δ such that:

$$0 < \Delta < \pi/2$$

For the point P (passive), we notice by drawing a parallel line to OP through W that:

$$2\beta_p = \phi_p + \Delta$$

For the point A (active), we notice by drawing a parallel line to OW through A that:

$$\pi - 2\beta_a + \phi_p = \Delta$$

Let:

$$2\beta_p - \phi_p = \Delta$$

$$2\beta_a - \phi_p = \pi - \Delta$$

This is in accordance with the fact that $2\beta - \phi_p$ is defined by its sine.

Some authors, somewhat arbitrarily, state that the direction whose image is closest to O, that is whose image is the point A, corresponds to the active state meanwhile the point P corresponds to the passive state. This terminology stems from the fact that for the image A, the horizontal compression stress on the vertical wall is less than the vertical stress on a horizontal plane whose image is A' and, in fact, the active state under a free horizontal surface is such that the major principal stress is vertical.

We see that, for the point P (supposedly passive state), the stress on the wall is more intense than the vertical stress, but this would no longer be true if P was to the left of the point X the intersection of Mohr's circle and the perpendicular to 0σ at W.

Ultimately, and merely conventionally, we can say that the state is active if the minor principal stress σ_2 is close to the horizontal axis. The image A of the wall is close to σ_2 and consequently, the major principal stress is close to the vertical axis.

The state is said to be passive if σ_1 is close to the horizontal axis.

1.5. Measurement of the mechanical properties of divided solids

1.5.1. Consolidation stress

Let us consider a cylindrical sample initially subjected to a consolidation stress σ_c on all its sides. This pressure is indispensable if we want the sample not to collapse. Thus, the flour escaping from a burst bag spreads almost like a liquid. It does not have rigidity. But if we squeeze the same flour in the hand, it takes the shape of the fingers and the palm. It has thus acquired rigidity and we say that it is consolidated.

In the measuring instruments used, the three principal stresses are not equal and the definition of the consolidation stress depends on the instrument used.

1.5.2. Measurement of displacements as a function of stresses

Röck and Schwedes [RÖC 05] describe a "uniaxial" cell as well as how to make use of it. Schwedes and Schulze [SCH 90] describe the "biaxial" cell whose implementation requires much time and is thus reserved to research laboratories. Haaker and Rademacher [HAA 83] describe an equipment said to be "triaxial".

The sample is placed in a rubber cylinder of circular cross-section and closed at both ends using a metallic circle. One of the circles acts as a piston on which we exert a thrust, which compresses the sample of the D S, which fills the cylinder. On the lateral surface, the intermediate principal stress σ_2 is equal to the minor principal stress σ_3.

The pressure σ_1 (major principal stress) is exerted by the piston on the sample. We establish the relationship between the piston's displacement and the pressure σ_1.

The stresses σ_2 and σ_3 are equal to the pressure exerted by the external fluid on the cylinder's rubber wall. A device enables the measurement of the lateral displacements ε_2 and ε_3.

This device enables the measurement of σ_1 and $\sigma_2 = \sigma_3$ (pressure of fluid).

We call deviatoric stress the difference $\sigma_1 - \sigma_3$ and we usually say that:

$$q = \left(\sigma_1 - \sigma_3\right) / 2$$

We call average pressure:

– in two dimensions: $p = (\sigma_1 + \sigma_3)/2$;

– in three dimensions: $p = (\sigma_1 + \sigma_2 + \sigma_3)/3$

The variables p and q are essentially used only in soil mechanics. As we know, the implementation of these solids in chemical industries, food industries, etc. involves pressures of the order of 0.5 bars at most, whereas soil mechanics acts on pressure up to 50 bar.

Roscoe *et al.* [ROS 58] made use of this equipment to establish the existence of the curve of critical porosities. This curve is a skew curve (nonplanar) in the p, q, e coordinate system where $e = \dfrac{\varepsilon}{1-\varepsilon}$ (ε here is the porosity of the D.S.).

This triaxial apparatus is only utilized in soil mechanics.

1.5.3. *Measurement of the limit shear stress for shear displacement [SCH 90]*

The reference apparatus for this measurement is the said Jenike's cell named after its inventor. Jenike *et al.* [JEN 60] give a detailed description of the cell as well as how to make use of it. We find additional information in Kurz's article [KUR 76].

Carr and Walker [CAR 67] suggest and describe an annular cell, which, contrary to that of Jenike, enables a shear displacement at constant speed in time and of idenfinite duration.

The "Hang-up Indicizer" of Johanson is described by Bell *et al.* [BEL 94] gives, according to its authors, results which do not always coincide with those of Jenike's cell.

NOTE.–

Luong [LUO 93] adopts an original logic based on the study of the hysteresis phenomenon on rice and wheat.

NOTE.–

Authors have reviewed the different ways of defining divided solids. We have: Eisenhart-Roth and Peschl [EIS 77], Schwedes and Schulze [SCH 90], Schwedes [SCH 96], Schwedes [SCH 00], and Ganesan *et al.* [GAN 08].

NOTE.–

Ganesan *et al.* [GAN 08] review the parameters affecting the flowability and the ease of handling divided solids taking as example grains subjected to drying.

1.5.4. Measurement of the surface area and the true (intrinsic) mass density of a divided solid

Ergun [ERG 51] suggested a method using the percolation of a gas through the D.S. for measuring the volume area of a D.S.

We shall not emphasize on the liquid displacement method, which is very classical and easy to use for the measurement of mass density.

1.6. Stockpiling

1.6.1. Angle of repose (also called "natural embankment angle")

We will acknowledge the expression given by Sokolowski [SOK 65]:

$$tg\alpha = \frac{\left[2\exp\left[(\pi - 2\alpha)tg\phi\right] - 1\right]tg\phi}{2\left[\exp\left[(\pi - 2\alpha)tg\phi\right] - 1\right]tg\phi}$$

α: angle of repose (rad)

ϕ: internal friction angle of the D.S. (rad)

The angle of repose is the largest angle, which the D.S.'s free surface can make with the horizontal plan.

Strictly speaking, this calculation is only applicable to the upper part of the pile.

EXAMPLE 1.1.–

$\phi = 20°$ $\tan \phi = 0.364$

Let us assume that $\alpha = 30° = 0.523$ rad.

$$\text{tg}\alpha = \frac{\left[2\exp\left[0.364\times(3.1416-2\times0,523)\right]-1\right]0.364}{2\left[\exp0.364(3.1416-2\times0,523)-1\right]-0.364^2} = 0.59159$$

We obtain:

$$\alpha = 30.6° \,\#\, 30°$$

Our approximation was correct.

NOTE.– Influencing parameters of the angle of repose.

The system's geometry comes into picture. Thus, in increasing order of the angle of repose:

– conical pile;

– slope whose ridge is rectilinear. An example is the case that we have studied above;

– circular crater in the form of a cone pointing downward;

– circular pits. On this matter, we shall refer to the work of Jenike and Yen [JEN 62b].

According to the nature of the considered solid, still in increasing order of the angle of repose:

– smooth spheres (steel balls);

– smooth ellipsoid;

– rough sphere;

– angular solid.

The finer the particles are the larger the angle of repose is, as the mass subjected to gravity decreases faster than its surface.

1.6.2. Practical values of the angle of repose

The following table provides some orders of magnitude. It is partly established based on the results obtained by Brown and Richards [BRO 65].

Nature of the solid	Angle of repose
Hard and smooth unimodal granular spheres $\bar{d}_p > 500\mu m$	$\alpha_E < 15°$
Rough or angular particles of low sphericity. Unimodal granular $60_{\mu m} < \bar{d}_p < 300\mu m$	$25° < \alpha_E < 40°$
Bimodal mixture (fine with granular, hence compactness). Between 10 and 30% of fine solid $\bar{d}_{p1} < 60\mu m$ $\bar{d}_{p2} >> 60 \mu m$	$\alpha_E > 50°$
Consolidated powder (interparticular attraction) unimodal granular $\bar{d}_p < 60\mu m$	$\alpha_E > 80°$

Table 1.4. Angles of repose

Ileleji and Zhou [ILE 08] illustrate the measurement of the angle of repose of some grains. Moreover, Adamiec et al. [ADA 05] state a relationship between the angle of repose and the internal friction angle.

1.6.3. Importance of pile storage

Pile storage is required for a volume greater than or equal to 20,000 m^3. The product can be either in the open air, or kept in a container which shall be:

– either only covered, if we want to shelter the product from the rain;

– or at the same time covered and closed on the sides if we are looking to protect the product not only from the rain but also from the dispersion of dust in the environment.

The conformation of the piles can be achieved with the help of tape distributors or more rarely, by a noria (unbroken succession) of trucks.

The recovering can be done, optionally, by:

– a bucket wheel. These wheels are movable;

– scrapping by bulldozers like on salt marshes. The usually irregular path followed by the bulldozers make the characteristics of the recovered product unpredictable;

– underground pit (tunnel) containing a conveyor belt. The horizontal dimensions of the pit should be sufficient to avoid the arching and/or the formation of pits in the mass covering the pit.

1.6.4. Importance of the study of free surfaces and underlying stresses

Minerals and some industrial products, like salt for application on roads, or coal before combustion are stored in bulk. The congestion of the piles thus produced depends on the product's angle of repose and the effective stresses in the product can lead to arching if the duration of storage is too prolonged.

1.6.5. Stresses on free surfaces

On every free surface, no matter its orientation, the shear is zero:

$$\tau = 0$$

So, one of the two principal directions of stresses is normal to this surface. But, given that we are dealing with a free surface, this principal stress is zero and in plane coordinates it can only be the minor principal stress σ_2.

$$\sigma_2 = 0$$

The major principal stress σ_1, parallel to the free surface, is less than or equal to the unconfined limit stress f_c:

$$2\bar{\sigma} = \sigma_1 \leq f_c$$

1.6.6. Principal stresses under a horizontal free surface

On the free surface, the boundary conditions are, referring to section 1.2.3:

$$\tau_{xy} = 0$$

$$\sigma_y = 0 \ (\text{no vertical overloading})$$

$$\sigma_x = \text{constant} = q \ (\text{uniformity across the entire plane})$$

We assume a uniform field of stresses, not only on the free surface, but also on the entire horizontal level, irrespective of its depth. The partial derivatives with respect to x are therefore nil and the equilibrium equations are written as:

$$\frac{\partial \tau_{xy}}{\partial y} = 0 \qquad \frac{\partial \sigma_y}{\partial y} = -\gamma$$

γ: specific gravity of the divided solid: $N.m^{-3}$

Let us now integrate taking the boundary conditions into consideration:

$$\sigma_y = -\gamma y \qquad \tau_{xy} = 0$$

The shear stress τ_{xy} being nil, the horizontal and vertical axes are principal directions for the stress matrix.

1.6.7. Active and passive states

When the depth is large enough or when the value of lateral compression σ_y exceeds a certain limit, Mohr's circle at every point comes in contact with the LRS (see Figure 1.22):

1) If $\sigma_x < \sigma_y = \gamma y$, the state of the stresses is said to be active and the average stress $\overline{\sigma}_a$ is then such that:

$$\overline{\sigma}_a + \sigma_o = \frac{\gamma y}{1 + \sin\phi} = \frac{\sigma_x}{1 - \sin\phi}$$

$$\sigma_o + \overline{\sigma}_a = \frac{\sigma_x + \gamma y}{2}$$

Φ is obviously the -static internal friction angle.

In the event of breakage, the soil goes further downward under the influence of $\sigma_y = \gamma y$.

2) If $\sigma_x > \sigma_y = \gamma y$, the state of the stresses is said to be passive and, in this state, the average stress $\bar{\sigma}_p$ is such that:

$$\bar{\sigma}_p + \sigma_0 = \frac{\sigma_x}{1+\sin\phi} = \frac{\gamma y}{1-\sin\phi}$$

$$\sigma_0 + \bar{\sigma}_p = \frac{\sigma_x + \gamma y}{2}$$

In the event of breakage, the soil rises by lateral compression.

While the value of the vertical stress σ_y is prescribed by the specific gravity γ and the depth y, the lateral stress σ_x remains a free parameter in this problem and thus the average stress $\bar{\sigma}$ is minimum in the active state and maximum in the passive state (see Figures 1.21 and 1.22).

So, if σ_x is nil and when Mohr's circle is in contact with the LRS, the state is inevitably active.

However, along the surface σ_y is nil and the state is either undefined if σ_x is nil, or passive. It is only from a certain depth that σ_y can exceed σ_x and the state can become active.

Finally, let's recall that the sliding planes to breakage always make the angle $\varepsilon = \pm\left(\frac{\pi}{4} + \frac{\phi}{2}\right)$ with the direction of the major principal stress σ_1 (Figures 1.21 and 1.22).

In fact, the sliding is normal at the point of static breakage.

In the active state σ_1 is vertical,

In the passive state σ_1 is horizontal.

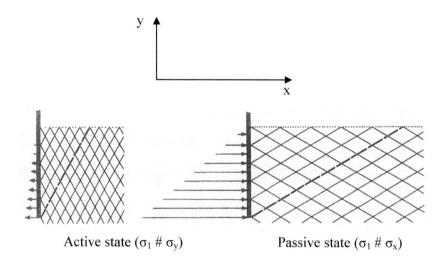

Active state ($\sigma_1 \# \sigma_y$)   Passive state ($\sigma_1 \# \sigma_x$)

Figure 1.21. *Orientation of the limiting*
slicklines according to Wieghardt [WIE 75]

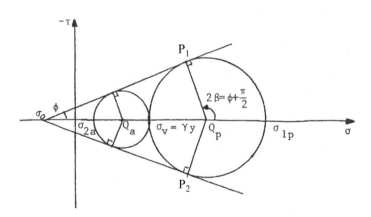

Figure 1.22. *Mohr's circles*

The circle with center Q_a corresponds to the active state and, for this circle:

$$\sigma_1^{(a)} = \sigma_y$$

The circle with center Q_p describes the passive state and, for this circle:

$$\sigma_1^{(p)} = \sigma_x$$

In Figure 1.22, the angle 2β formed by the $Q_p\sigma_{1p}$ axis with Q_pP_1 is:

$$2\beta_1 = \phi + \frac{\pi}{2} \quad \text{for the point } P_1 \text{ (active)}$$

For the point P_2 (passive):

$$2\beta_2 = -\varnothing + \frac{3\pi}{2}$$

The two planes subjected to σ_x are normal to these directions. So the angle α formed by these two planes with the major principal stress axis σ_{1P} is:

$$\alpha_1 = \frac{\pi}{2} - \beta_1 = \frac{\pi}{4} - \frac{\phi}{2}$$

$$\alpha_2 = \frac{\pi}{2} - \beta_2 = -\frac{\pi}{4} + \frac{\varphi}{2}$$

Let:

$$\alpha = \pm\left(\frac{\pi}{4} - \frac{\phi}{2}\right) = \pm\varepsilon$$

The + sign corresponds to the passive state.

EXAMPLE 1.2.–

Charcoal is stored in bulk and forms a rectangular pile for which the two vertical sections are trapezoidal. Knowing that the height of the top plane with respect to the ground is 4 m, find the major principal compression stress σ_1 at ground level.

The horizontal stress is almost nil, and hence the state is of active type and the stress σ_1 is the "hydrostatic" pressure.

For charcoal:

$$\rho_a = 950 \text{ kg.m}^{-3}$$

$$\gamma = 9.81 \times 950 = 9320 \text{ N.m}^{-3}$$

$$\sigma_1 = \gamma h = 4 \times 9320 = 37\,280 \text{ Pa} = 0.37 \text{ bar}$$

2

Stresses in Hoppers and Silos: Filling, Emptying and Content Homogeneity

2.1. Stresses on the walls

2.1.1. The two states of stress

The state of the stresses (active or passive) depends on the orientation of the major principal stress σ_1 (see Figure 2.1):

– if σ_1 is close to the vertical axis, that is to the weight of the product, the state is said to be active as it is this weight that brings about the stresses quite independently. The active state is obtained by loading the hopper with closed emptying;

– if σ_1 is close to the horizontal axis, the state is said to be passive and corresponds to the flowing product. But a flowing product slightly expands and exerts a force on the lateral walls. So the major principal stress is close to the horizontal axis. Thus, the flowing product "passively" experiences the stresses acting as a result of the lateral walls.

When emptying a silo or hopper, the D.S. initially appears to be in the active state. As soon as the discharge is released, the solid switches to the passive state moving from the opening and going upwards. However, the change of state stops at the cylinder–cone junction such that the cylinder remains in the active state.

When the drainage process is stopped, the solid remains in its passive state. Conversely, during the first fill of a hopper, the state is active (σ_1 is vertical) from top to bottom of the hopper.

From the point of view of the force exerted on the walls, we will see that the passive state is more severe than the active state.

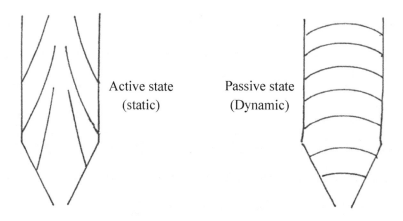

Active state   Passive state
(static)   (Dynamic)

Figure 2.1. *Surface lines of the major principal stress* σ_1

2.1.2. *Resulting sign conventions*

We will be looking at the representation of the *right part of the section of hoppers and silos*. As for the angles, the positive direction of rotation is the counterclockwise direction.

Let $\bar{n}$ be the unit vector of the normal to the wall and *entering inside the product* in the hopper. The compression stress σ exerted by the wall on the product will be counted positively in the direction of $\bar{n}$.

Now let $d\bar{n}/d\theta$ be the unit vector derived by $\dfrac{\pi}{2}$ in the counterclockwise direction. The shear stress τ exerted by the wall on the product makes up for the acting gravity and is directed upward opposite to the vector $d\bar{n}/d\theta$ in whose direction the stress $-\tau$ will be acting. The algebraic measure of τ will therefore be negative and obviously $-\tau$ will be positive (see Figure 2.2).

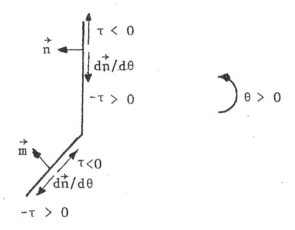

Figure 2.2. *Supporting unit vectors to the stresses acting on the wall of a hopper*

With these conventions, from Mohr's representation, τ is negative and $-\tau$ is positive. The image of the stresses is therefore in the first quadrant.

2.1.3. Janssen's equations [JAN 95] for a vertical cylindrical wall

The equation is derived assuming that the friction on the wall is totally involved, either in the active state, or the passive state.

Let h be the depth, measured in the product from the free surface, $\overline{\sigma}_v$ the average value of the vertical compression stress on the area and τ_p the shear stress acting on the wall.

The vertical pressure force acting on an area A at depth h is acting downward and equal to $A\overline{\sigma}_v$.

This force increases with the volume Ω of the product present above the area A: if h increases by dh, this volume increases by $d\Omega$ of which the weight will be $\gamma d\Omega$.

But the reaction force of the vertical lateral walls acts upward and against the weight. This reaction force increases by $\tau_p dS$ where dS is the lateral surface area.

Finally, the equilibrium equation is:

$$A d\overline{\sigma}_v = \gamma d\Omega - \tau_p dS$$

γ: specific gravity of the stored product: $N.m^{-3}$

But:

$$d\Omega = Adh \qquad dS = Pdh$$

P: perimeter of the section (m).

In addition, let's say:

$$B = \frac{\tau_p}{\sigma_v}; \qquad D = \frac{\sigma_v}{\sigma_v} \qquad \text{where: } BD = \frac{\tau_p}{\sigma_v}$$

σ_v is the vertical stress acting on the right side of the wall.

Let us integrate the equilibrium equation after replacing τ_p with its expression in terms of $\overline{\sigma}_v$.

$$\overline{\sigma}_v = \frac{A\gamma}{PBD}\left[1 - \exp\left[-\frac{PBDh}{A}\right]\right] + \overline{\sigma}_{vs}\exp\left[-\frac{PBDh}{A}\right]$$

$\overline{\sigma}_{vs}$: overload stress R on the free surface of the product (Pa).

The coefficient of friction on the wall is $\tan\phi_p$ and the friction is entirely involved. Hence the expression of the horizontal compression stress σ_h on the vertical wall:

$$\sigma_h = \frac{\tau_p}{\tan\phi_p} = \frac{BD\overline{\sigma}_v}{\tan\phi_p}$$

NOTE.–

According to a more common formulation, Janssen's equation is written as:

$$\overline{\sigma}_v = \frac{A\gamma}{P\lambda\mu}\left[1 - \exp\left[-\frac{P\lambda\mu h}{A}\right]\right]$$

The coefficient of friction here is given by:

$$\mu = \tan\phi_p = \tau_p/\sigma_h$$

The factor, which we call "Janssen's ratio", is:

$$\lambda = \frac{\sigma_h}{\overline{\sigma}_v}$$

Hence:

$$\sigma_h = \frac{A\gamma}{P\mu}\left[1 - \exp\left[-\frac{P\lambda\mu h}{A}\right]\right]$$

Furthermore, we see that:

$$\frac{\tau_p}{\overline{\sigma}_v} = \lambda\mu = BD$$

The ratios λ and μ are, however, considered empirical (DIN 1055 standard) whereas the ratios B and D are calculated based on the value of the angle ϕ_p of friction on the wall and based on the value of the dynamic internal friction angle δ.

The calculation of the distribution rate D of the vertical stress is done by Walters [WAL 73a, WAL 73b] for a circular cross-section. For this purpose, we assume that the shear stress varies linearly from the axis of the section up to the wall.

2.1.4. Average stress on the cylindrical wall

The average stress $\overline{\sigma}_{cy}$ on the wall is defined by:

$$\overline{\sigma}_{cy} = \frac{\sigma_1 + \sigma_2}{2}$$

It is the half-sum of the principal stresses.

Therefore, if the state is assumed to be active:

$$\overline{\sigma}_{cy} = \frac{\sigma_h}{1 + \sin\delta\cos2\beta_a} = \frac{A\gamma}{P_a\tan\phi_p}\left[\frac{1 - \exp\left[-\dfrac{PBDh}{A}\right]}{1 + \sin\delta\cos2\beta_a}\right]$$

2.1.5. Expression of parameters according to the state of stresses (cylindrical section)

The cylinder is assumed to be of circular cross-section.

1) Coefficient B:

This coefficient is defined by:

$$B = \frac{\tau}{\sigma_v}$$

τ: shear stress on the wall (Pa)

σ_v: vertical compression stress (Pa)

This coefficient B is given by:

Walters' equation 7 [WAL 73a]

Walters' equation 12 [WAL 73b].

In these equations, the + sign corresponds to the active state and the − sign to the passive state.

2) Coefficient D:

This coefficient is defined by:

$$D = \frac{\sigma_v}{\overline{\sigma}_v}$$

$\overline{\sigma}_v$: average vertical compression stress on the horizontal section (Pa).

The coefficient D is given by:

Walters' equation 3 [WAL 73a]

Walters' equation 5 [WAL 73b].

The + sign corresponds to the active state and the − sign corresponds to the passive state.

3) Product BD:

$$BD = \frac{\tau}{\overline{\sigma}_v}$$

The product BD is given by:

Walters' equation 13 [WAL 73a].

2.1.6. Supporting reaction exerted by the wall (cylindrical part)

Assuming the wall friction is completely utilized, the reaction of the cylindrical wall is given by:

$$R_{Scy} = \int_0^h \tau_p dS$$

The product located *below the depth h sustains an average vertical pressure* $\overline{\sigma}_v$ corresponding to the force $A\overline{\sigma}_v$.

The weight of the solid $\gamma\Omega_h$ from the surface to the depth h is balanced by:

– the reaction of the cylindrical wall;

– the reaction of the product $A\overline{\sigma}_v$.

Finally:

$$\gamma\Omega = R_{Scy} + A\overline{\sigma}_v$$

2.1.7. Study of the convergent: simplifying assumption

Let's assume, as previously, that the friction on the wall is completely involved:

$$\tau_p = \sigma_p \tan\phi_p$$

If $\vec{n}$ is the normal unit vector to the wall, we have:

$$\left(0\overline{\sigma}_1, \vec{n}\right) = \beta$$

β is the angle corresponding to the active or passive mode.

We will assume that the fact that the wall is inclined does not considerably change the ratio:

$$BD = \frac{\tau_p}{\overline{\sigma}_v}$$

(Nonetheless Walters [WAL 73b] provides a graphical method to take this inclination into consideration, provided it is low).

The equilibrium equation:

Let us assume the convergent is full up to the depth h.

The total height of the convergent is H.

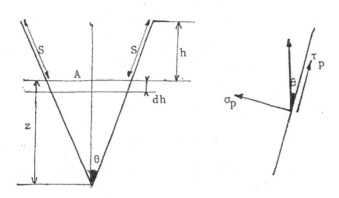

Figure 2.3. *Equilibrium of the convergent*

Let us go beneath the free surface. The vertical pressure force acting on a cross-sectional area A of the convergent located at the height z acts downward and is equal to:

$$A\overline{\sigma}_v$$

This force increases with the volume Ω of the product located above the cross-sectional area A. If h increases by dh, the weight of this volume will increase by:

$$\gamma d\Omega$$

But the reaction of the lateral walls acts upwards, and against this weight:

$$t_{pv} dS$$

The equilibrium equation is written as:

$$d\left(A\overline{\sigma}_v\right) = \gamma d\Omega - t_{pv} dS \qquad [2.1]$$

or:

$$A\frac{d\overline{\sigma}_v}{dh} + \overline{\sigma}_v \frac{dA}{dh} + t_{pv}\frac{dS}{dh} - \gamma\frac{d\Omega}{dh} = 0$$

Let us determine the terms of this equation:

$$A = \pi (H - h)^2 \tan^2\theta$$

$$S = \pi \frac{\tan\theta}{\cos\theta} \left(H^2 - z^2 \right) \qquad\qquad (h = H - z)$$

$$\frac{1}{A}\frac{dS}{dh} = -\frac{1}{A}\frac{dS}{dz} = \frac{2}{z\cos\theta} = \frac{2}{(H-h)\sin\theta} > 0$$

$$\frac{1}{A}\frac{dA}{dh} = \frac{-2}{H-h} < 0$$

$$t_{pv} = \tau_p \cos\theta + \sigma_p \sin\theta = \tau_p \left[\cos\theta + \frac{\sin\theta}{\tan\phi_p} \right]$$

$$t_{pv} = \tau_p \frac{\sin(\theta + \phi_p)}{\sin\phi_p} = \overline{\sigma}_v BD \frac{\sin(\theta + \phi_p)}{\sin\phi_p}$$

$$\Omega = \frac{\pi}{3} \left[H^3 - (H - h)^3 \right] \tan^2\theta$$

$$\frac{d\Omega}{dh} = \pi (H - h)^2 \tan^2\theta = A$$

Hence:

$$\frac{d\overline{\sigma}_v}{dh} - \overline{\sigma}_v \times \frac{2}{H-h} + \overline{\sigma}_v \times \frac{2 \times BD \sin(\theta + \phi_p)}{(H-h)\sin\theta \sin\phi_p} - \gamma = 0$$

Let's set:

$$K = 2 \left[\frac{BD \sin(\theta + \phi_p)}{\sin\theta \sin\phi_p} - 1 \right]$$

We can deduce that:

$$\frac{d\overline{\sigma}_v}{dh} + \frac{K\overline{\sigma}_v}{H-h} - \gamma = 0$$

Let us now perform a change of variables:

$$h = H - z \quad \text{and} \quad dh = -dz$$

The equilibrium equation is finally written as:

$$\frac{d\overline{\sigma}_v}{dz} - \frac{K}{z}\overline{\sigma}_v + \gamma = 0$$

In order to solve this equation, let's say:

$$\overline{\sigma}_v = vz \; ; \; \text{hence} : \frac{d\overline{\sigma}_v}{dz} = v + z\frac{dv}{dz}$$

Finally, Walters [WAL 73b] shows that after integration the equilibrium equation is written as (Walker's equation 21, [WAL 66]):

$$\overline{\sigma}_v = \frac{\gamma z}{K-1}\left[1 - \left(\frac{z}{z_0}\right)^{K-1}\right] + \sigma_{vo}\left(\frac{z}{z_0}\right)^K$$

The cylindrical part acts like an overload with respect to the conical part. This overload will be defined by the index s.

Changing the notation, we will write:

$$\text{for } z_0 = z_s \quad \overline{\sigma}_v = \overline{\sigma}_{vs}$$

Hence, for $z < z_s$:

$$\overline{\sigma}_v = \frac{\gamma z}{K-1}\left[1 - \left[\frac{z}{z_s}\right]^{K-1}\right] + \overline{\sigma}_{vs}\left[\frac{z}{z_s}\right]^K$$

If z_s is equal to H, the overload is simply caused by the weight of the product present in the cylindrical part.

In order for the stress $\overline{\sigma}_v$ to be zero, in accordance with experimental results for $h = 0$, it is necessary that:

$$K > 0 \quad \text{that is} \quad \theta < \theta^* = \text{Arctg}\left[\frac{BD\,\text{tg}\phi_p}{\text{tg}\phi_p - BD}\right]$$

But, depending on whether the state of the stresses is active or passive:

$$(BD)_A \ll (BD)_P \tag{2.2}$$

$$K_A \ll K_P \tag{2.3}$$

$$\overline{\sigma}_{vA} \gg \overline{\sigma}_{vP} \tag{2.4}$$

In fact, $\overline{\sigma}_v$ is therefore a decreasing function of the parameter K, which in itself is an increasing function of BD.

The upper limit of θ defined by θ^* below is then an existence criterion for the active state in the convergent.

The compression stress on the wall is, in general:

$$\sigma_p = \frac{BD}{\tan\phi_p}\overline{\sigma}_v$$

When the overload $\overline{\sigma}_{vs}$ is zero, and depending on the depth h, σ_p can be maximum either for the active state or the passive state, as shown in the Figure 2.4.

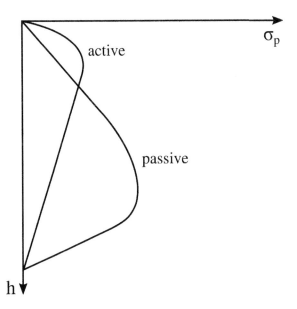

Figure 2.4. *Stress profile on the wall without overload*

2.1.8. Supporting reaction exerted by the wall (convergent)

This reaction results from the vertical component of the stress exerted on the product by the convergent's inclined wall:

– the frictional grip $\tau_p \cos\theta$;

– the vertical reaction of the pressure $\sigma_p \sin\theta$.

If the friction is completely involved:

$$\sigma_p = \frac{\tau_p}{\tan\phi_p}$$

Hence:

$$t_{pv} = \tau_p \left(\cos\theta + \frac{\sin\theta}{\tan\phi_p} \right) = \tau_p \frac{\sin\left(\theta + \phi_p\right)}{\sin\phi_p}$$

The resulting force for the entire lateral surface S of the convergent is:

$$R_{co} = \int_0^{h=H} t_{pv} dS$$

The bracketing force R_{co} is zero at the upper level of the convergent and the previous integral therefore has to be calculated moving from the free surface towards the bottom:

$$R_{co} = \int_{h=0}^{h=H} t_{pv} dS$$

But from the equilibrium equation:

$$R_{co} = \int_0^h t_{pv} dS = \gamma \int_0^h d\Omega - \int_0^h d(A\bar{\sigma}_v) = \gamma\Omega_{co} + A_s\bar{\sigma}_{vs} - A\bar{\sigma}_v$$

The term $-A\bar{\sigma}_v$ corresponds to the support by the inferior product, that is located under the plane of side z:

For h = 0 $A\bar{\sigma}_{vs} = A_s\bar{\sigma}_{vs}$ (free surface)

For h = H $A\bar{\sigma}_v = 0$ (tip of the convergent)

Thus for the convergent assembly, the wall balances the weight of the product *and the overload's force* $\bar{\sigma}_{vs}$.

2.1.9. *Rotation of the principal stresses to the cylinder–convergent transition*

If we go from the cylinder to the convergent, the direction of the major principal stress makes a sharp angular turn of Γ (which Enstad called Δ [ENS 77]). Using Chasles' equation, we can write:

$$\Gamma = \left(\overrightarrow{\sigma_{cy}}, \overrightarrow{\sigma_{lco}}\right) = \left(\overrightarrow{\sigma_{lcy}}, \vec{n}\right) + \left(\vec{n}, \vec{m}\right) + \left(\vec{m}, \overrightarrow{\sigma_{lco}}\right)$$

$\vec{n}$ and $\vec{m}$ are respectively the unit vectors entering the product concerning the cylinder and the convergent. If θ is the half-angle at the top of the convergent:

$$(\vec{m},\vec{n}) = \theta > 0$$

Hence:

$$\Gamma = \beta_{cy} - \beta_{co} - \theta$$

We are referring to Enstad's equations 6 and B5 [ENS 75]. This author demonstrates how to find Γ.

2.1.10. Inclined wall

Let us consider an elementary thickness of the product having the shape:

– of an arch if the cross-section of the convergent is rectangular;

– or of a dome if the cross-section is circular.

The system is considered to be in the passive state, and the angle made with the horizontal plan of the arch surface (or of the dome) is $(\beta_p + \theta)$. In fact, the major principal stress is parallel to the arch surface (or of the dome).

Enstad stated that the thrust ΔF_1 from bottom to top due to the peripheral supports balances the weight ΔW of the elementary layer of the layer of material, as well as the force of magnitude ΔF_2 due to the variation of the vertical compression stress.

This translates into Enstad's differential equation 9 [ENS 77]:

$$r\frac{d\overline{\sigma}}{dr} - X\overline{\sigma} = -\gamma X r$$

where $\overline{\sigma}$ is the average stress:

$$\overline{\sigma} = \frac{\sigma_1 + \sigma_2}{2}$$

Where r is the distance between the point in question and the top of the convergent.

The solution of the differential equation is in the form:

$$\overline{\sigma}(r) = \frac{\gamma Y r}{X-1} + \left[\overline{\sigma}(R) - \frac{\gamma Y R}{X-1}\right]\left[\frac{r}{R}\right]^X$$

where R is the distance between (along the wall) the top of the convergent and the cylinder–convergent boundary.

This equation is given by Enstad's equations 8 and C8 [ENS 75] and by Enstad's equation 12 [ENS 77].

Let's say:

$$a = \frac{\gamma Y}{X-1} \quad \text{and} \quad b = \overline{\sigma}(R) - \frac{\gamma Y R}{X-1}$$

$$\overline{\sigma}(R) = \overline{\sigma}_{co} = \rho\overline{\sigma}_{cy}$$

where R is the length of the convergent generator m.

Finally, $\overline{\sigma}(r)$ is in the form:

$$\overline{\sigma}(r) = ar + b\left[\frac{r}{R}\right]^X$$

Hence:

$$\frac{d\overline{\sigma}}{dr} = a + \frac{bX}{R^X}r^{X-1}$$

The maximum value of $\overline{\sigma}$ is obtained for:

$$r_M = \left[\frac{a}{-bX}\right]^{\frac{1}{X-1}} R^{\frac{X}{X-1}}$$

The maximum is important for the capacity of the wall of the cone.

The expression of the parameters X and Y is given by:

1) Active state:

 – circular cross-section: Enstad's equations 13 and 14 [ENS 77];

 – rectangular cross-section: Enstad's equations 11 and 12 [ENS 75].

2) Passive state:

 – circular cross-section: Enstad's equations 10 and 11 [ENS 77];

 – rectangular cross-section: Enstad's equations 9 and 10 [ENS 75].

2.1.11. "Holding" reaction of the product on the walls of the convergent

We shall use Enstad's equations for stresses:

$$R_{co} = \int_S t_{pv} dS = \int_0^R \frac{\sin(\theta + \phi_p)}{\sin\phi_p} \tau_p dS$$

where θ is the half-angle at the top of the convergent,

 ϕ_p is the friction angle on the wall, and

 τ_p is given by Enstad's equation, which is of the form:

$$\tau_p(r) = \sin\delta \, \sin 2\beta \bar{\sigma}(r)$$

Let:

$$\tau_p(r) = \sin\delta \, \sin 2\beta \left[ar + b \left[\frac{r}{R} \right]^X \right]$$

where r is the distance from the top of the convergent m.

Since:

$$dS = 2\pi r \sin\theta \, dr$$

$$R_{co} = K.I$$

$$\text{where: } I = \int_0^R \left[ar + b\left[\frac{r}{R}\right]^X \right] r \, dr = \frac{aR^3}{3} + \frac{bR^2}{X+2}$$

$$K = 2\pi \frac{\sin\left(\theta + \phi_p\right)}{\sin\phi_p} \sin\theta \, \sin\delta \, \sin 2\beta$$

where δ is the angle made by the $o\sigma_1$ axis with the dynamic rupture line.

The value of the angle β and the expressions of a, b, and X differ depending on whether the active or passive state prevails in the convergent. The passive state corresponds to flow and persists when the flow stops. The active state only corresponds to initial loading before any emptying is done.

Naturally, the weight of the solid has to be compensated by the sum of the "holding forces":

$$R_{cy} + R_{co} = \text{Weight of the mass contained in the silo}$$

2.1.12. Existence of the active state in the convergent

We have two almost similar conditions:

1) that which arises from the cancellation of the stresses at the tip of the convergent:

$$\text{for BD} < \tan\phi_p \; : \; \theta < \theta^* \quad \text{with} \quad \theta^* = \text{Are tg} \left[\frac{BD \, \text{tg}\phi_p}{\text{tg} \, \phi_p - BD} \right]$$

2) that which was also derived by Enstad from the cancellation of the stresses at the tip of the convergent:

$$\theta < \theta^* \quad \text{with} \quad \theta^* = \frac{\pi}{2} - \beta_a$$

EXAMPLE 2.1.–

Consider a hopper of circular cross-section:

$D_T = 4$ m $h = 6$ m $\theta_c = 15°$

$\Delta = 45°$ $\phi = 20°$ $\phi_p = 15°$

$\gamma = 11772$ N.m^{-3}

For completely involved friction forces on the wall:

$\Delta = 21.47°$

$\beta_a = 86.76°$

$\beta_p = 18.23°$

Table 2.1 compares the results obtained by the methods of Walters and Enstad respectively:

	R_{cy} (passive)	R_{co} (cyl. passive)	R_{cy} (active)	R_{co} (cyl. active)
Walters	774, 443	481, 165	112, 474	1, 143, 134
Enstad	774, 443	475, 159	112, 474	1, 007, 063

Table 2.1. *Stresses on the wall of a cylindro-conical silo*

This table calls for the following remarks:

1) Walters' method is approximate in that the factor BD is derived from arbitrary assumptions, but our method of calculating R_{co} is such that we strictly have:

$$R_{co} + R_{cy} = M_E g$$

with M_E being the mass contained in the silo.

2) Enstad's equations, as he admitted himself, are only approximate and this explains the gaps which exist between the values of the vertical reaction R_{co} of the convergent. These gaps nonetheless remain less than 15%.

NOTE.–

If the reader is interested in the stresses not on the wall but within the D.S., he can refer to the following publications:

Lakshman Rao and Venkateswarlu [LAK 75]

Delaplaine [DEL 56].

2.1.13. Weakness of the wall to the cone-cylinder junction

We will show that the pressure on the wall increases sharply when we cross this junction moving from the cylinder to the cone.

But

$$Z = z/d \quad \text{and} \quad \overline{S}_z = \frac{\overline{\sigma}_z}{\rho g d}$$

Walters' equation (19) [WAL 73b] becomes *for the cone*:

$$\overline{\sigma}_z = \rho g \frac{(d - 2z tg\theta)}{2 tg\theta (K-1)} \left[1 - \left(\frac{d - 2z tg\theta}{d - 2z_h tg\theta} \right)^{K-1} \right] + \overline{\sigma}_{zo} \left(\frac{d - 2z tg\theta}{d - 2z_o tg\theta} \right)^K$$

z is oriented vertically upwards from the tip of the cone.

At the point z_0 of the cylinder–cone boundary:

$$d = 2z_o tg\theta \text{ and we verify that } \overline{\sigma}_z = \overline{\sigma}_{zo}$$

On the cylinder side, Walters' equation 14 [WAL 73a] becomes:

$$\overline{\sigma}_z = \frac{\rho g d}{4BD} \left(1 - \exp\left(\frac{-4BDz}{d} \right) \right) + \overline{\sigma}_{zo} \exp\left(\frac{-4BDz}{d} \right)$$

z is oriented vertically downwards from the top of the cylinder.

At the cylinder–cone boundary:

$$z = 0 \text{ and we verify that } \overline{\sigma}_z = \overline{\sigma}_{zo}$$

According to Walters' equations (2) and (5) (or 13 and 15) [WAL 66], the shear stress τ_p on the wall is in terms of B and D.

$$\tau_p = B\tau_{zp} = BD\bar{\sigma}_z \quad \text{but} \quad \tau_p = \tau_{rp}\tan\phi$$

So the stress τ_{rp} is normal to the wall and we deduce that:

$$\tau_{rp} = \frac{BD\bar{\sigma}_z}{tg\theta}$$

When emptying, the cylinder is in the active state and the cone is in the passive state. In the changeover plane, the value of BD changes from $(BD)_A$ in the cylinder to $(BD)_P$ in the cone and $(BD)_P$ is much greater than $(BD)_A$ as we can see from Walters's Figure 6 [WAL 73a].

Let's recall these paradoxical definitions:

– the passive state is a state of motion (dynamic);

– the active state is a state of rest (static).

Mohr's circle helps in understanding the increase in the stress on the wall when we move from the cylinder (which is in the active state) to the cone (which is in the passive state).

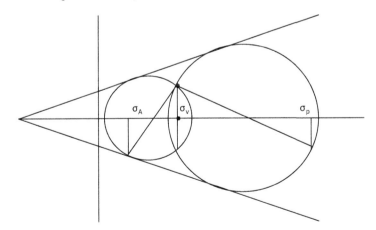

Figure 2.5. *Transition from the active state to the passive state*

We see that

$$\sigma_p \gg \sigma_A$$

However, we have not taken the inclination along the vertical wall of the cone into consideration. In order to do this, we will take reference from Enstad [ENS 75].

Note that Takahashi *et al.* [TAK 79] wrongly add the pressure stresses acting on two different surfaces in order to take the silo's fault on the cylinder–cone junction into consideration (their equation 30). This so-called justification of a so-called peak pressure is wrong.

Jenike and Johanson [JEN 68] show that, during the drain of a silo, the cylindrical part remains in the active state and the conical part changes to the passive state. They show that the changeover occurs at the level of the cylinder–cone junction (page 118 of their publication).

Jenike and Johanson [JEN 69b] came up with a simple expression for the extreme pressure P_{max} on the wall, which is attained at the level of changeover (their equation (19)).

NOTE.–

Hoppe and von Eisenhart-Rothe [HOP 80] directly measured the stresses in the products within the silo.

2.1.14. Weakness in the previous considerations

The fact that the friction on the wall is not completely involved everywhere leads to indeterminacy of the stresses inside the cylinder (which can actually explain the uneven results obtained from some measurements).

This indeterminacy is, as we shall see, related to the heterogeneity of the product contained within the silo and this heterogeneity itself results from the charging mode, as well as the nature of the product.

We can therefore say that:

$$\sigma = \alpha\sigma_A + (1-\alpha)\sigma_p \quad 0 < \alpha < 1$$

Hence:

$$\sigma_P < \sigma < \sigma_A \quad \left(\text{and not } \sigma = \sigma_A\right)$$

α here is the volume fraction of the product in the active state.

This will reduce the stress on the wall at the point where the changeover occurs.

But, from what we already know, we do not know how to find α clearly and precisely.

This method requiring us to set $\alpha = 1$ is therefore a precautionary measure.

2.2. Variation in the stresses with ensiling and desiling. Homogeneity, heterogeneity of content

2.2.1. Homogeneity and flowability of silage

The Reimberts [REI 71] talk about the experiments they conducted with plant seeds or the sand with which they fill a cylinder.

These authors noticed that as soon as the loading is complete, there is settling and the level of the content drops. On the other hand, the force on the bottom of the cylinder reduces and this can be explained by considering the fact that the state of the stresses which, immediately after the loading were completely of active type, changes progressively and partially to the passive type.

The products used by these authors were very fluent. In other words, their flowability $C = 1/f_c$ was very high and their flow ratio k_E was likely greater than 30. But we know that if we fill to maximum capacity with a clear liquid, as a liquid possesses an infinite flowability, the volume thus occupied will most likely be homogeneous. On the other hand, if we carry out the same operation with a divided solid without taking specific precautions, we will observe variations in the size of the particles and the porosity immediately after the filling.

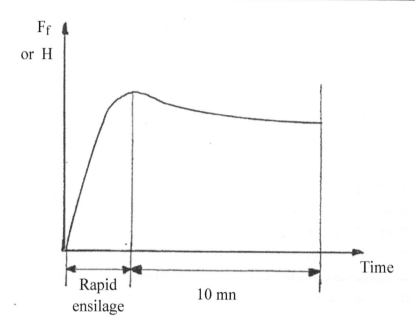

Figure 2.6. *Rate of change of the ensiled average level and of the force exerted on the flat bottom*

If the flowability $C = 1/f_c$ of the product is low, these heterogeneities will continue to exist. On the contrary, if the product is very fluent, we would observe within the 10 min following the filling that the product has settled with rearrangement and modification of the mutual positions of the particles. This change is due to gravity and contributes to the reduction of the corresponding potential energy.

This supports the remark made by Jenike and Johanson according to which if the extractor located beneath the hopper is relatively flexible, it experiences during loading a deflection and a subsidence due to the increasing load. The authors, in fact, recommend a low discharge rate during loading so that the pressure stresses on the wall, maximum in the passive state, progressively start acting immediately after loading and counter the vertical force exerted on the extractor.

If this precaution is taken, the loaded capacity would be completely in the active state and when the discharge system will be activated, the rapid change to the passive state along the height of the hopper could pose risks in regard to the load resistance of the equipment.

2.2.2. Importance of the properties of the product

Some products with high flowability include plant seeds and dry sand. Generally, we're referring to products of low or zero cohesion and also of low internal friction angle. In fact, these particles usually have a high sphericity and a smooth surface.

The less fluent products are usually chemically synthesized crystals and crushed or raw ores. In fact, the particles of these products are angular and rough and they have low sphericity.

The compressibility of the products can become involved but only when the vertical displacement due to packing becomes high. But in the cylindrical section of a hopper, this displacement is made as a whole, that is without deformation and can therefore only slightly help in the rearrangement of the particles. We can, however, expect the compressibility to be less negligible inside the convergent.

The powders ($d_p < 60$ or 100 μm) are significantly less fluent if they are outgassed and compact and, on the contrary, perfectly fluent if they are airy.

The average bulk density of a product is maximum if the product is homogeneous as there is therefore an absence of "caverns" or cavities.

2.2.3. Filling modes

Although it is paradoxical, a description of the emptying begins by a description of the filling mode, because the filling ("silage") controls the homogeneity of the ensiled load.

Essentially, two methods can be employed for the loading process:

– in a uniform rain across the area. This method is slow but guarantees maximum homogeneity.

– in a jet (generally at a constant speed). This method, being quick, creates high heterogeneities in the loaded mass.

On the contrary, in the rain method, the apparent density is maximum and draws closer to the one obtained by vibrations. On the contrary, the jet method allows the existence of low density regions such that the average density is reduced.

2.2.4. Characteristics of loading of a capacity

We distinguish:

1) the degree of consolidation (resulting from the value of the major principal stress) which defines the size of the rigidity domain. The consolidation increases from top to bottom of the silo or the hopper and also with the height of fall during loading by jet. The apparent mass density increases as consolidation increases;

2) the active or passive state of the stresses which defines the direction of the minor and major principal stresses. After a first fill, the state is still active, but in the convergent it generally becomes passive after a partial drain and remains passive even during a halt irrespective of the duration;

3) the value of the coefficient of friction $\tan \phi_p$ on the wall which decreases slightly when the normal stress on the wall increases;

4) the state (plastic or rigid) of the product depending on whether or not the intensity of the stresses is sufficient to reach the static rupture line (LRS);

5) the homogeneity of the product. After filling by jet, the state and the characteristics of the product differ from one area to another within the product and also on the wall.

2.2.5. Filling by jet (of granules) and localized over pressures

During the fall of the particles within a jet inside a hopper, the air resistance is negligible if we are dealing with granules and not powders. The speed acquired by the granules on arriving is given by:

$$V_f = \sqrt{2gh}$$

If the jet possesses the traverse section A on reaching the product at rest and if the flow rate of the mass of solid particles is W_S, the pressure exerted on the product will be:

$$P = \frac{F}{A} = \frac{W_S V_f}{A} = \frac{W_S \sqrt{2gh}}{A}$$

EXAMPLE 2.2.–

$$W_S = 100 \text{ kg.s}^{-1} \quad A = 0.1 \text{ m}^2 \quad h = 10 \text{ m}$$

$$P = \frac{100\sqrt{2 \times 9.81 \times 10}}{0.1} = 14,000 \text{ Pa}$$

$$P = 0.14 \text{ bar}$$

If the mass density of the product is 1,200 kg m^{-3}, this pressure is equivalent to a height H of the product equal to:

$$H = \frac{14,000}{1,200 \times 9.81} = 1.2 \text{ m}$$

2.2.6. Homogeneous ensiling and heterogeneous ensiling

For grains, the height of the fall needs to be minimized.

For powders, the ensiling process should be slow enough to allow for proper de-aeration. But the homogeneity thus achieved would come at the cost of loss of flowability.

In the homogeneous silage, the mass density is increased. But, all the stresses are proportional to this mass and if the other parameters are kept constant (ensiled height and state of the stresses), the stresses would be greater if the mass is homogeneous, in particular, the force F_f exerted on the bottom surface. But one should not forget that for a homogeneous D.S., the ensiled mass will be greater at equal height.

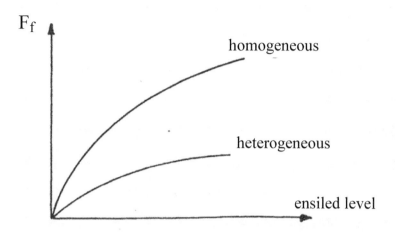

Figure 2.7. *Force exerted on the bottom surface following the method of silage*

The mass density of the product is minimum for a heterogeneous load. In fact, due to the segregation of the fine and the granules, the fine particles no longer fill the gaps existing between the grained particles and the global porosity increases.

The scale of heterogeneities is proportional to the size of the granules and if this magnitude is greater than $1/10$ of the diameter D_T of the hopper, we will be able to witness fluctuations in the flow rate or of the stresses during extraction. That is to be expected for a small hopper containing granules.

2.2.7. Homogeneous ensiled mass (slow and rainwise filling)

The apparent density as well as the consolidation of the product are at their maximum values. There is no segregation.

If the wall is rough, it is possible that the product, especially if it has been closely packed (by vibrations for example) "drops off" the wall in bulk and thus leads to a dangerous impact exactly during the opening of the drainage. It is then a question of the free fall of a large mass of the product.

Whether there was an initial impact or not, the stresses will subsequently reduce progressively and smoothly during the discharge process (desilage).

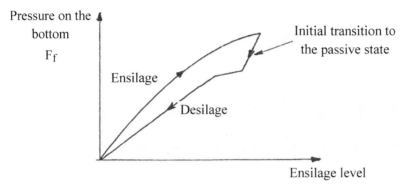

Figure 2.8. *Ensilage-desilage cycle for a homogeneous load*

During the desilage, there can be a changeover to the passive state from bottom to top of the silo. If the bottom is flat, the passive state can appear, if it was not pre-existing, but we notice that the increase in the thrusts on the walls can only be transmitted partially because the pre-existing active state could only be partial or, in the stationary product, an effect of cushioning "pads" actually appears.

2.2.8. Heterogeneous ensiled mass (quick fill and, for the granules, in a jet)

Due to the homogeneity of apparent density, there exist high consolidation regions and low consolidation regions.

If the wall is smooth, a "dropout" suddenly appears at the start. On the other hand, if the wall is rough, the vertical stress reduces globally during emptying, but this reduction occurs in fits and starts.

When the flow starts, the changeover to the passive state cannot happen simultaneously throughout the mass. In fact, the scope of an "arch" cannot equal the size of the silo as, due to heterogeneity, some regions of weakness are always present. Thus, only local changeovers to the passive or state can exist.

The parietal arching leads to the momentary braking of the descent, of the product located above the arch. But the inferior product continues to descend and the less supported arch collapses, the hook on the wall reduces and the pressure on the bottom increases.

The product then finds itself being localized and packed and a new parietal arch can appear, which increases the hook on the wall and causes a drop in the pressure on the bottom.

It follows that the hook up on the wall of a heterogeneous and fluctuating mass is altogether less than if the homogeneity was greater. The force on the bottom will then have to be greater.

Ultimately, as Reimbert and Reimbert [REI 71] have observed, the force on the bottom of the drain follows a broken line made of three different types of line segments (see Figure 2.9):

1) end of "archings",

2) formation of new "archings", and

3) simple reduction in the ensiled volume due to the progress of the drain.

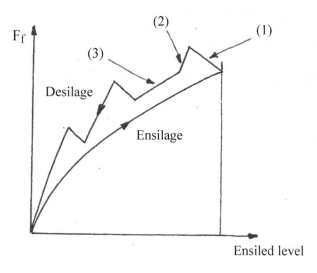

Figure 2.9. Ensilage-desilage cycle of a heterogeneous mass

Jenike, observing the flow in a hopper having transparent walls, had even noticed the formation of momentarily stationary and irregular patches in contact with the wall, and he had on the other hand noticed the existence of a parietal pressure peak at the same point. His interpretation was that of a local changeover from the active state to the passive state.

3

Draining of Hoppers and Silos: Stresses and Flow Rate

3.1. General information

3.1.1. Specific investment

The necessary investment for storage increases in the following order:

– stockpiling;

– shop for bags, barrels, and drums;

– silo or hopper.

The relative investment on a capacity increases in the following order depending on the chosen material:

– aluminum;

– mild steel; and

– stainless steel.

The investment varies with the effective volume V and it is defined by the specific investment I_V (currency unit/m^3):

– if V < 200 m^3, the construction is done in a workshop and remains economical even considering transportation to the site. I_V decreases with V;

– if V > 200 m^3, transportation is no longer possible and production has to be done on the construction site. I_V then increases with V.

The limit 200 m^3 is attained for transportation by road or by rail.

Once installed and equipped, the material has its cost doubled from the factory price, that is the manufacturing price.

3.1.2. Methods of ensiling (of capacity filling)

The unloading time of a 30 ton truck is about 30 min, but if the unloaded solid has to be transported for 200 m, the unloading time can increase to 2 h or more. Right out of the factory, the ensiling time varies from 5 min to 2 h at most.

Ensiling can:

– in a testing station, be realized in the form of rain distributed across the cross-section;

– make use of gravity (jet from a mechanical conveyor) if direct contact with the atmosphere is without drawback;

– be pneumatic. Arrival in the capacity can be tangential to ease (centrifugally) the separation of the solid from air, but it can also be directed downwards, if the product is not too dusty. A bag filter contains dust particles in the upper portion of the capacity and air is sent out to the atmosphere by a fan. A depacking device can be mounted on the lower portion (conical) of the silo or hopper.

A high level of the product leads to a halt by default and the supply is then interrupted.

3.1.3. Instruments for the measurement of the level of the product

1) Capacitive sensor: We record the dielectric capacitance between the sensor and the wall (the properties of the medium are modified by the presence of the product).

2) Ultrasounds: We can, through the measurement of their absorption, continuously measure the level of the product.

3) Radioactive radiation: This also helps in the continuous measurement of the level of the product, but the installation is costly and difficult.

4) Fixing the capacity on load cells, which would give the measure of the weight of the product.

5) Vibrating plate whose frequency changes if it is immersed in the product. Such a device can only be used as level alarm (up or down).

3.1.4. What is a hopper or a silo?

It is generally a closed capacity designed for the storage of a divided solid. We suggest that we distinguish:

– hoppers whose total volume is less than about 100 m³;

– silos whose total volume is greater than 100 m³.

θ is the half-angle at the tip of the convergent. More concisely, we will call this the angle of inclination (implied: with respect to the vertical axis).

The withdrawal opening can be:

– either circular of diameter D_o;

– or slot-shaped. A slot is a rectangle whose length L_o is equal to at least three times its width ℓ_o.

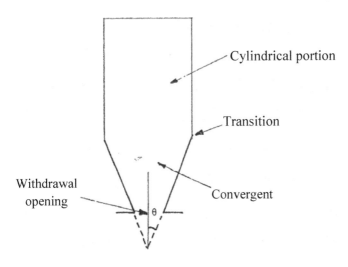

Figure 3.1. *Storage capacity of a divided solid*

The "size" B_o of the opening is, depending on its shape:

either $B_o = D_o$ or $B_o = \ell_o$

The cross-section of the cylinder can be:

– circular;

– rectangular; and

– more rarely (due to mechanical reasons) square.

A circular opening generally corresponds to a circular or square cylinder and a rectangular cross-section naturally corresponds to an opening in the form of a slot. However, given the advantages of this last type of opening, it can be associated to a cylinder of circular or square cross-section.

3.1.5. Silos with multiple outlets

We can equip the flat bottom of a silo with multiple outlets. We can thus control the withdrawal rate by commissioning only the necessary number of tappings. On the other hand, this design creates space in height with respect to a unique convergent and, if the product is very fluent (like cereals or plant seeds), the flow will be in bulk without wells or arching. However, if the open outlets are eccentric, a bending force will be exerted on the cylindrical portion of the silo and could be damaging if the slenderness (ratio of height to width) of the silo is high.

3.2. Flow types and flow regimes

3.2.1. Types of flow

Traditionally, we distinguish, depending on the design of the hopper or the silo:

– block flow during which the product is not deformed. The descent of a divided solid in the cylindrical portion of a hopper is done in this way if the product slides like a piston in the cylinder;

– mass flow during which the entire product present (all its mass) in the hopper is in motion. The product is deformed (in the convergent this deformation is inevitable);

– flow with "perforation" of the solid mass by a well or a trench concerns only the product close to the element of symmetry (axis or plane) of the hopper. This product descends in a well while, between the well or the trench and the wall, a dead zone contains only stationary solid. As the level goes down, the walls of the well or the trench, which are exposed, collapse and the hopper is emptied in its entirety. However, the steady vertical height is proportional to the cohesion of the product and thus, the hopper will not be emptied completely.

3.2.2. Mass flow regimes depending on the product

We can distinguish between three regimes:

1) for the granules without cohesion, the rate of flow is constant and regular.

Above the cylinder–convergent transition point (e.g. at a vertical distance of the order of a diameter), there is the formation of a crater drowned in the mass of the product, which means that the product speeds up its descent close to the axis or the plane of symmetry.

If the angle at the top of the convergent is large (and *a fortiori*, for a flat bottom surface) between the well or the trench and the wall, the fall of the product is almost nil. The walls of the well are slightly inclined to the vertical axis and converge more strongly close to the outlet. Under the crater, there is rotation of the principal directions of the stress tensor, which leads to the expansion of the product and a reduction in its apparent mass density.

2) for the granules with cohesion (sawdust), the rate in extreme circumstances can be reduced to a series of successive crumblings.

Oblique faults are formed on the walls of the convergent and spread firstly toward the top then, more horizontally, towards the axis. They can reach the axis above the cylinder-convergent transition point. Above these faults, the product is in the "active" state whereas, below them, it has changed to the "passive" state and expanded (see Figure 2.1); the average stress has therefore reduced. It has become plastic and can also go down in the convergent.

Along the descent, the faults move slowly with the product such that entire domains comprised between two faults keep their individuality excluding a deformation. On the upper part of the convergent, new faults are formed as replacements to those which have descended.

It is the successive arrival at the opening ("crumbling") of the domains comprised between two faults which can give the flowrate and the stresses on the wall their fluctuating property.

3) for the fine products (d_p < 300 μm), we shall see that the permeability of air is low and that strongly reduces the flowrate of the solid. This phenomenon aggravates if the product is compressed, which reduces its porosity. Thus, according to Bulsara et al. [BUL 64], barium carbonate can be sucked but not "forced".

NOTE.–

In the previous cases (1) and (2), when the descending product reaches either the cylinder-convergent transition point or the level of the crater, it changes state from the active state to the passive state, and enters an "agitated" zone where the particles undergo intense movements, which gives rise to a characteristic sound level. If the flow rate is fluctuating, we are dealing with cracking (formation of faults) and if the flowrate is constant, we perceive rustling (rearrangement of particles and expansion).

The top of the agitated zone is at a higher level than the cylinder–convergent transition point. This level:

– is proportional to the width of the silo;

– increases with the thickness of the opening; and

– increases with internal friction.

The product then leaves the agitated zone and enters the zone called the "free fall" zone in the immediate vicinity of the outlet. The separation between agitated zone and free fall zone is progressive and imperceptible for granules. On the other hand, the separation is clear for a fine product.

3.2.3. Flow character according to the configuration of the convergent

We distinguish:

1) Mass flow (without dead zone):

The flow is said to be as a whole when the entirety of the product present is in motion and is somewhat deformed. For this, the half-angle θ at the top of the convergent should not go above an upper limit θ_{Ma} or θ_{Mp} depending on the state of the stresses:

If $\theta < \theta_{Ma}$ the stresses in the mass flowing product are in the active state (major principal stress σ_1 close to the vertical axis).

If $\theta_{Ma} < \theta < \theta_{Mp}$ the stresses in the still mass flowing product are in the passive state.

2) Flow with a dead zone:

If $\theta > \theta_{Mp}$ the flow occurs in a "convergent" whose half-angle at the top is close to θ_{Mp}. The walls of this "convergent" are made up of solids at rest.

Depending on the roughness of the wall of the convergent part of the hopper, the dead zone is momentary or permanent.

3) Flow with perforation:

A well or a trench of vertical walls exists in the middle of the stationary solid mass and, "perforates" it. Periodically, the upper portion of the well collapses and falls apart in the latter. The flowrate is therefore chaotic.

The occurrence of perforation is determined not only by the geometric properties of the container (hopper or silo) but also by the properties of the solid (flowability).

3.3. Criteria for mass flow

3.3.1. Importance of capacities without dead zones

In such capacities, for hoppers or silos, the renewable volume (active volume) identifies with the entire volume of the ensiled solid. These capacities are distinguished by the following two advantages:

– Withdrawal flow is of the "en masse" type.

– Complete drainage is always possible and even easy to carry out.

The dwelling time is approximately the same for all the particles between complete emptying and filling (first in = first out). Thus, every kind of segregation is avoided and, if the product is to change over time, it would be easy to control this change for the entire load by limiting the time in the hopper or silo.

3.3.2. Criteria for mass flow (active state of the stresses)

Enstad's parameter X [ENS 77] must be positive for the stresses, as experience shows, to cancel out at the top of the convergent.

But, in the active type, the parameter X is:

$$X = \frac{m \sin\delta}{1 + \sin\delta} \left[\frac{\sin(2\beta_a + \theta)}{\sin\theta} - 1 \right]$$

(Enstad's equations 13 and 14, [ENS 77]).

So an active state will only be possible if:

$$\sin(\theta + 2\beta_a) > \sin\theta$$

or:

$$\cos\theta \, \sin 2\beta_a > \sin\theta (1 - \cos 2\beta_a)$$

$$\cos\beta_a \, \cos\theta - \sin\beta_a \, \sin\theta > 0$$

$$\cos(\theta + \beta_a) > 0$$

$$-\frac{\pi}{2} < \theta + \beta_a < \frac{\pi}{2}$$

$$-\beta_a - \frac{\pi}{2} < \theta < \frac{\pi}{2} - \beta_a$$

But, we know that (see section 1.4.4):

$$\beta_a = \frac{\pi}{2} - \frac{1}{2}(\Delta - \phi_p)$$

ϕ_p is the friction angle on the wall.

Hence:

$$\theta_{Ma} < \frac{1}{2}(\Delta - \phi_p)$$

With:

$$\sin\Delta = \frac{\sin\phi_p}{\sin\delta} \quad (\text{see section } 1.4.4)$$

3.3.3. Limited active type domain – angle of approach

If the overall flow in the active state is not possible, then the centre of the product could still be in the active state but, close to the wall, the product will remain stationary.

The surface separating the flowing product and the stationary product constitutes a convergent whose angle at the summit is the angle of approach β.

Due to the fact that the product does not slide on the wall anymore but on itself, ϕ_p becomes equal to ϕ (internal friction angle) and, as ϕ is greater than ϕ_p, the friction consequently increases.

Thus, the flow occurs in the limited domain where the active state of stresses prevails. This domain has the shape of a coaxial cone to the convergent and whose angle of approach will be determined by the equation:

$$\beta = \frac{1}{2}(\Delta_o - \phi) \quad \text{with } \sin\Delta_o = \frac{\sin\phi}{\sin\delta}$$

Let:

$$\beta = \frac{1}{2}\text{Arcsin}\left[\frac{\sin\phi}{\sin\delta}\right] - \frac{\phi}{2}$$

EXAMPLE 3.1.–

$$\phi = 20° \qquad \delta = 45° \qquad \theta = 15°$$

$$\Delta_o = \text{Arcsin}\left[\frac{0.3420}{0.7071}\right] = 28.92 \text{ sexigesimal degrees}$$

$$\beta = \frac{1}{2}(28.92 - 20) = 4.45° < \theta$$

The flow corresponds to the formation of a well on the axis of the hopper. At first, a depression (crater) appeared on the surface of the product.

3.3.4. Conditions for mass flow (passive state of the stresses)

Let us assume plane coordinates. According to our conventions, if we place $-\tau$ on the ordinate axis in Mohr's representation, this quantity is positive, and on the other hand, the orientation of the angles is retained.

We will translate the two hypotheses below for a point M of the wall:

– an overall displacement of the product exists along the inclined wall of the convergent;

– simultaneously, the product is *not supposed to be able to* slide on itself. It follows from these two hypotheses that the major principal stress σ_1 does not show discontinuity in any direction close to the wall contrary to what happens at the boundary between cylinder and convergent of a hopper.

If $\vec{v}$ is the downward vertical unit vector and $\vec{g}$ is the unit vector dependent on the displacement of the sliding product, the flow will only be possible if the oriented angle $\vec{g}, \vec{v}$ is positive.

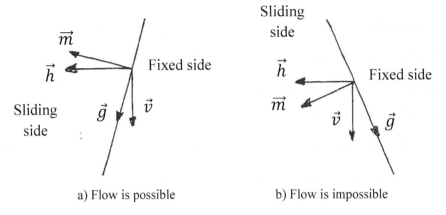

a) Flow is possible   b) Flow is impossible

Figure 3.2. *Flow criteria*

If we have:

$$\left(\vec{g},\, \vec{m}\right) = -\frac{\pi}{2}; \quad \left(\vec{h},\, \vec{v}\right) = +\frac{\pi}{2}$$

$$\left(\vec{g},\, \vec{v}\right) = \left(\vec{g},\, \vec{m}\right) + \left(\vec{m},\, \vec{h}\right) + \left(\vec{h},\, \vec{v}\right) = -\frac{\pi}{2} + \left(\vec{m},\, \vec{h}\right) + \frac{\pi}{2}$$

$$\left(\vec{g},\, \vec{v}\right) = \left(\vec{m},\, \vec{h}\right)$$

And the previous condition becomes, for the flow to be *impossible*:

$$\left(\vec{m},\, \vec{h}\right) < 0$$

The slide on the wall is, anyway, completely involved and, for the passive state (see Figure 3.3):

$$2\left(\vec{\sigma}_1,\, \vec{m}\right) = 2\overrightarrow{\beta_p} = \Delta + \phi_p$$

Moreover:

$$\left(\vec{n},\, \vec{h}\right) = \theta$$

Hence:

$$\left(\vec{\sigma}_1,\ \vec{h}\right)=\left(\vec{\sigma}_1,\ \vec{n}\right)+\left(\vec{n},\ \vec{h}\right)=\theta+\frac{1}{2}\left(\Delta+\phi_p\right)$$

But:

$$2\beta_S=2\left(\vec{\sigma}_1,\ \vec{m}\right)=\phi+\frac{\pi}{2}$$

Hence:

$$\left(\vec{m},\ \vec{h}\right)=\left(\vec{m},\ \vec{\sigma}_1\right)+\left(\vec{\sigma}_1,\ \vec{h}\right)=\theta+\frac{1}{2}\left(\Delta+\phi_p\right)-\frac{1}{2}\left(\phi+\frac{\pi}{2}\right)$$

And $\left(\vec{m},\ \vec{h}\right)<0$ is equivalent to:

$$\theta<\theta_{Mp}$$

With:

$$\theta_{Mp}=\frac{1}{2}\left(\frac{\pi}{2}+\phi-\Delta-\phi_p\right)$$

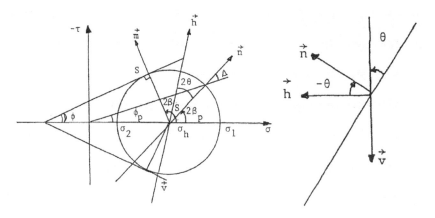

Figure 3.3. *Mohr's circle*

3.3.5. Jenike's criteria [JEN 87] for the existence of a desilage well (emptying)

In 1964 and 1965, Jenike [JEN 64a, JEN 64b, JEN 65] included the equilibrium equations for the calculation of the radial stress in a convergent. He then realized that his calculation could be improved and in 1987 [JEN 87] he solved the system of equations comprising of the equilibrium equations and the equation of conservation of matter (its equation said to be that of continuity). It results in the equation of the radial velocity (his equation 26).

After integration, Jenike's Figure 6 [JEN 87] divides the first quadrant with a curve. The friction angle on the wall lies on the ordinate axis and the angle made by the wall with the vertical axis lies on the abscissa.

Close to the origin, the flow occurs en masse. Above the curve, it is the flow with perforation that is predominant.

Actually, Jenike [JEN 87] suggests two curves:

– one for a hopper of horizontal circular cross-section;

– the other for an elongated rectangular cross-section

Jenike [JEN 62a, JEN 62b] investigates the stability of a well with axial symmetry and gives the profile of the stationary D.S after the flow of a part of the wall of the well.

EXAMPLE 3.2.–

Product:	ϕ	ϕ_p	Δ	θ_{Ma}	θ_{Mp}	θ (Jenike) (plane coordinates)
Sand	31	16.5	33.5	8.5	35.5	32
Polystyrene	39	14.5	23.4	4.4	45.5	33
Glass beads	25	16.5	42.2	12.8	28.2	32

We notice a corresponding change between the values of θ_{Mp} and those proposed by Jenike in plane coordinates.

3.4. Flow with dead zone

3.4.1. *Dead zones: features and disadvantages*

If the angle θ crosses the limit θ_{Mp}, a dead zone appears between the wall and the withdrawal opening.

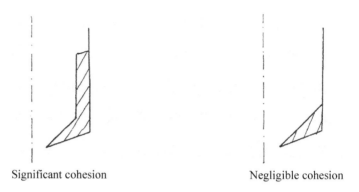

Significant cohesion   Negligible cohesion

Figure 3.4. *Principal profile of the dead zone (wide hopper)*

The vertical height of the dead zone increases proportionately with the cohesion of the product – at the expense of the renewable volume. So the more a product possesses cohesion, the more the volume of the sacrificed capacity in the dead zone increases. If the cohesion is nil, the internal surface of the dead zone reduces to that of a crater whose angle with the horizontal is approximately equal to the angle of repose.

The presence of a dead zone has the following drawbacks:

– as we have just seen, a reduction in the net volume, that is the renewable volume;

– difficult precise changes in the dead volume and, consequently, of the net volume;

– slow change always possible and even the degradation of the product with time in the stationary zone;

– during complete emptying, the derived product can turn out to be heterogeneous either because it has changed in the dead volume, or because the heterogeneity was created in the course of loading jetwise.

In fact, during the loading in a jet, the product constitutes a heap on the slopes of which the granules flow easily and appear on the periphery while the fine products percolate vertically across the grains and find themselves at the center of the hopper or silo.

3.4.2. Conditions for complete drainage – adherence in the convergent

When the level of the moving product drops, an almost vertical wall of stationary product is left exposed and if the cohesion of the product is not very high, that wall can collapse. On the other hand, if the wall of the convergent is not too rough and if it is not too distant from the vertical axis, the hopper could be emptied immediately.

At the end of an emptying process, the residual blocks of product remaining on the wall of the convergent should be removed easily by sliding. This is the classical problem of the inclined plane and we know that sliding is only possible if the angle of slope α of the surface on the horizontal axis is greater than the friction angle ϕ_p^* of the product on the wall:

$$\alpha = \frac{\pi}{2} - \theta > \phi_p^*$$

Here we will assume the quasi-static value of ϕ_p^*, which is often $5°–10°$ greater than the dynamic value ϕ_p that we generally use in the determination of stresses on the wall. We shall write:

$$\phi_p^* = \phi_p + 10°$$

Ultimately:

$$\theta < 80° - \phi_p \qquad\qquad [3.1]$$

Jenike proposes:

$$\theta < 80° - \text{angle of repose.}$$

If we admit that the angle of repose is close to $(\phi_p + 5°)$, we are close to the condition [3.1].

The adherence to the wall leads to the hooking up of blocks, which can collapse without warning. The adherence:

– increases if the wall is rough, as roughness increases the number of hooking sites;

– results from a physicochemical adsorption by a wall of the product or some of its components. This adsorption increases with the duration of contact and hence that of storage.

In order to reduce and even avoid adherence, it usually suffices to coat the wall of the convergent with an elastomer or a plastic material (resin, epoxyresin) because these products are at the same time smooth and nonadsorbent.

3.4.3. Edge's slope with interior corners

We usually notice the permanent presence of products in the interior corners of convergents of square or rectangular cross-section. This ensures that the slope on the horizontal axis of the dihedral edges thus formed is much less than that of the sides of the said dihedrals.

Applying Pythagoras's theorem to the Figure 3.5:

$$OC^2 = OA^2 + OB^2 = CB^2 + CA^2$$

$$\frac{h^2}{\tan^2\theta} = \frac{h^2}{\tan^2\alpha} + \frac{h^2}{\tan^2\beta}$$

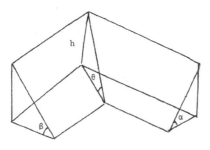
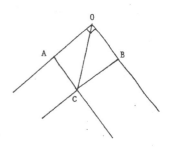

Figure 3.5. *Slope of the edges of the convergents of non-circular cross-section*

EXAMPLE 3.3.–

$\alpha = 45°$ $\beta = 54°$

$$\frac{1}{\tan^2\theta} = \frac{1}{1} + \frac{1}{1.894}$$

$\tan\theta = 0.809$

$\theta = 38.90°$

$38.9° < 45°$ and $38.9° << 54°$

3.4.4. Capacities with permanent dead zones

These are capacities for which:

– either the convergent is of a large angle;

– or the bottom surface is just flat.

Such heights reduce the height of the product's fall at the start of jetwise ensiling. In fact, this height includes that of the convergent B/(2tgθ). It follows that:

– the wear of the bottom surface due to the shocks is reduced;

– the pressure and the additional packing due to the gravitational energy dissipated during the shock upon arrival are also reduced. The cohesion happens to be reduced and the flowability improved.

During withdrawals, the product slides on the dead zone, which avoids the abrasion and the wear of the walls.

Finally, if the bottom surface is really flat, its price is clearly less than that of a conical bottom surface.

Permanent dead zone capacities are widely used for granules, which, as we know, have a flowability good enough for the dead volume to be reduced to a minimum. *A fortiori*, the large pieces ($d_p > 0.1$ m) falling on the bottom surface have less chances of damaging it.

The minerals possess these characteristics and as they have not yet been treated, the eventual segregation that they could experience does not pose any problem.

3.4.5. Operation of a permanent dead zone capacity without segregation

This type of capacity, which could have a flat bottom surface, is very elongated. In the lower portion of the capacity, there is a peripheral dead zone. On the other hand, in the upper portion of the cylinder, the flow is constant and occurs in a block.

If we agree to let the dead zone "sleep" indefinitely and to only partially utilize the capacity of the silo, we could have a homogeneous flow, provided we never go beneath 0.5 B (B is the diameter or width of the silo) above the cylinder–convergent transition point (or 1.5 B above the base if the bottom surface is flat).

3.4.6. Persistence of the dead zone during the flow

The solid in the stationary dead zone occupies zone (1) in the passive state and this zone's boundary corresponds to the direction of the unit vector $\vec{g}_1$ (see Figure 3.6). Along this boundary, the internal sliding of a solid is potentially possible without the stationary zone (1) being perturbed.

The flowing solid is in a state of active stresses and occupies zone (2) whose lateral boundary corresponds to the sliding direction of the unit vector $\vec{g}_2$.

Figure 3.6 corresponds to the case where the two zones are partially covered according to the common area which is hatched. In this case, the dead zone is eroded and progressively disappears during the flow. This configuration corresponds to:

$$\left(\vec{g}_1, \vec{g}_2\right) < \pi$$

On the other hand, if:

$$\left(\vec{g}_1, \vec{g}_2\right) > \pi$$

Then both zones are well separated as a dead angle exists between them. The dead zone can continue to exist and it is stable.

Let us translate these assumptions to Mohr's circle in Figure 3.7. At the extreme point E of the dead zone, both zones coexist:

– the zone (1) in the passive state;

– the zone (2) in the active state.

To each of these two states correspond respectively the circles of center W_p and W_a.

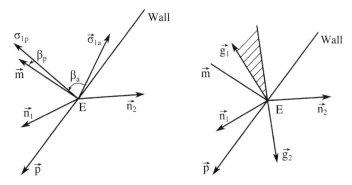

Figure 3.6. *Direction of the sliding surfaces $\vec{g}_1$ and $\vec{g}_2$ (dead zone)*

The image C of the normal to the wall corresponds to a slide fully involved for zone (2), in the active state where the solid is in motion. The coordinates σ and τ of the point C are therefore such that:

$$|\tau| = \sigma tg\phi_p$$

Emphasis is made on the fact that σ and τ are the normal and tangential stresses on the wall. These stresses are constant when we move from zone (2) to zone (1) as the state of the system is either balanced or slowly varying.

– Passive state side on the circle of center W_p:

$$2\beta_p = \Delta + \phi_p \quad \text{with} \quad \beta_p = \left(\vec{0\sigma}_{1p}, \vec{m}\right)$$

$\vec{m}$ is the unit vector normal to the wall.

The normal $\vec{n}$ to the internal sliding surface (1) has the point S as its image:

$$\left(\vec{0}\sigma_{1p}, \vec{n}_1\right) = \beta_S \quad \text{and} \quad 2\beta_S = \frac{\pi}{2} + \phi$$

– Side with active state, on the circle of center W_a:

$$\beta_a = \left(\vec{0}\sigma_a, \vec{m}\right) \quad \text{and} \quad 2\beta_a = \pi - \left(\Delta - \phi_p\right)$$

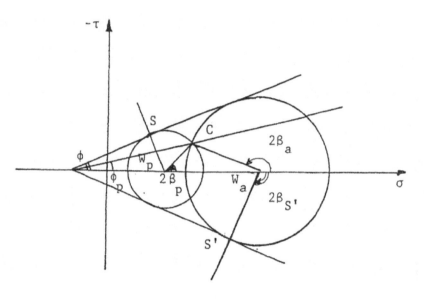

Figure 3.7. *Mohr's circles for the endpoint E (Figure 3.9) of the dead wall*

Let the vector $\vec{n}_2$ be normal to the internal sliding surface (2) with image S':

$$\left(\vec{0}\sigma_{1a}, \vec{n}_2\right) = \beta_{S'} \quad \text{and} \quad 2\beta_{S'} = -\left(\frac{\pi}{2} + \phi\right)$$

The angle of both sliding surfaces is identical to that of their normal vectors $\vec{n}_1$ and $\vec{n}_2$:

$$\left(\vec{n}_1,\ \vec{n}_2\right)=\left(\vec{n}_1,\overrightarrow{0\sigma_{1p}}\right)+\left(\overrightarrow{0\sigma_{1p}},\vec{m}\right)+\left(\vec{m},\overrightarrow{0\sigma_{1a}}\right)+\left(\overrightarrow{0\sigma_{1a}}+\vec{n}_2\right)$$

$$2\left(\vec{n}_1,\ \vec{n}_1\right)=-\left(\frac{\pi}{2}+\phi\right)+\beta_p-\beta_a-\left(\frac{\pi}{2}+\phi\right)$$

$$=-\pi-2\phi+\Delta+\phi_p-\pi+\left(\Delta-\phi_p\right)$$

$$\left(\vec{n}_1,\ \vec{n}_2\right)=-\pi+\left(\Delta-\phi\right)=\left(\vec{g}_1,\vec{g}_2\right)$$

Let us assume that the following inequality is true:

$$\left(\vec{g}_1,\ \vec{g}_2\right)=\left(\vec{n}_1,\vec{n}_2\right)<\pi$$

It translates into:

$$\phi>\Delta \quad \text{that is} \quad \sin^2\phi>\sin\phi_p$$

This means that the roughness of the wall is limited, the dead zone is unstable (it cannot get "get hooked" to the wall).

Conversely, the dead zone is permanent if the wall is sufficiently rough, that is if:

$$\sin\phi_p>\sin^2\phi$$

EXAMPLE 3.4.–

If $\phi=30°$:

$\sin\phi=0.5 \quad \sin^2\phi=0.25$

If $\phi_p>14.5°$

the dead zone will be permanent.

3.4.7. Practical data

Depending on the lifespan of the dead zone, we have:

1) Momentary dead zone:

At the start of withdrawal, all the walls and especially those of the convergent are in contact with some quantity of "hooked" and stationary product. Only a narrow stream of the product is flowing in the central area of the capacity. But as the withdrawal is performed, we notice a lateral expansion of the axial moving stream at the expense of the dead zone, which finally completely disappears (see Figure 3.8).

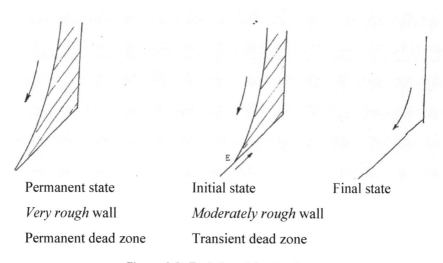

Permanent state   Initial state   Final state

Very rough wall   *Moderately rough* wall

Permanent dead zone   Transient dead zone

Figure 3.8. *Evolution of the dead zone*

2) Permanent dead zone:

We can observe an eventual initial lessening in the dead zone, but the latter ends up being stable.

In an axial opening, the distance s between the edge of the opening and the endpoint E of the dead zone increases as the verticality of the wall increases, which makes sense because the solid has more difficulty "hooking up" (see Nguyen *et al.* [NGU 80]).

$$\frac{s}{D_I} = 0.35 - 0.0412\theta$$

θ: half-angle at the top of the convergent: sexagesimal degrees

D_T: diameter of the hopper (m).

If the opening is in the form of a slot, some areas of weakness can arise locally along the slot such that the position of E appears to be fluctuating.

If the wall becomes rougher (ϕ_p increases), the distance s so reduces, consequently, as the surface of the wall in contact with the moving solid.

3.4.8. *Angle of approach and angle of convergence*

The convergence angle θ_{co} of the well or trench should not be mistaken for the angle of approach β (see Figure 3.9).

The point A is the common point between the vertical wall of the well and the wall of the convergent. The angle of the convergent θ is:

$$\theta_{co} = AOz$$

The curve AO along which the solid slides has a tangent OB at O and the corresponding angle is:

$$\gamma = BOz \quad (\gamma \text{ is the angle of approach})$$

We see that:

$$\theta < \gamma$$

θ is not greater than $15°$ and γ can be as high as $30°$.

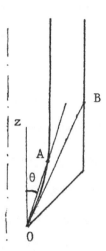

Figure 3.9. *Angle of approach and angle of convergence*

In their book, Brown and Richards [BRO 65, p. 156] suggest an estimation of γ based on the internal friction angle ϕ:

$$\gamma = 50° - \frac{\phi}{2}$$

3.4.9. Configuration of the well (angle of approach and diameter)

The results obtained by Giunta [GIU 69] give as a function of the dynamic internal friction angle:

– The angle of convergence θ_{co};

– A coefficient A (called "coefficient of flat bottom") which helps in the calculation of the width of the well.

The half-width of the well is therefore given by:

$$D_p = D_o + 2tg\theta_{co} \left[\frac{H - \dfrac{AD_o}{2}}{1 + Atg\theta_{co}} \right]$$

D_o: diameter of the bottom opening

D_p: diameter of the wells.

EXAMPLE 3.5.–

$$\delta = 40° \qquad D_o = 0.6 \text{ m} \qquad H = 8 \text{ m}$$

On Figure 2 of Giunta's book [GIU 69], we have:

$$\theta_{co} = 6° \qquad \tan\theta_{co} = 0.105 \qquad A = 2.2$$

Hence:

$$D_p = 0.5 + 2 \times 0.105 \left[\frac{8 - \dfrac{2.2 \times 0.6}{2}}{1 + 2.2 \times 0.105} \right]$$

$$D_p = 1.85 \text{ m}$$

NOTE.–

If the internal friction angle ϕ is, as it is often, in the neighborhood of $20°$, the angle of approach will be:

$$\gamma = \frac{1}{2} \left[\text{Arcsin} \left(\frac{\sin 20°}{\sin 40°} \right) - 20° \right] = 6.1°$$

This value is practically equal to the empirical value suggested by Giunta [GIU 69] for the convergence angle θ.

NOTE.–

Tüzün and Nedderman [TÜZ 82b] investigated the drainage of a grained type D.S. and confirmed the predictions made by their kinematic model.

3.5. Arching or doming and its prevention

3.5.1. *Minimum size of the discharge outlet*

Experience shows that the halt and interruption of the emptying of a hopper by the formation of an arch or a dome of product above the outlet are impossible if this outlet is sufficiently large, and whose size we will find by referring to Enstad's work [ENS 75, ENS 77].

The surface of an arch makes an angle α on the periphery, with the normal to the wall. But, the formation of arches or domes is an unpredictable phenomenon, which becomes significant when the vertical resultant of the peripheral thrust is maximum. Enstad shows that this happens when α becomes less than the value:

$$\alpha_M = \frac{\pi}{4} - \frac{\theta}{2}$$

The angle α will therefore have to be greater than this value if arching is to be avoided.

Another upper limit of α is the friction angle on the wall, ϕ_p. Above this value, sliding occurs. Ultimately, in order to prevent arching from occurring, it will suffice that:

if $\phi_p < \phi_M$ $\alpha > \phi_p$

if $\alpha_M < \phi_p$ $\alpha > \alpha_M$

On the other hand, the product should be found in a limiting internal state. In this case, the major principal stress is:

$$\sigma_1 = f_c + \frac{1+\sin\phi}{1-\sin\phi}\sigma_2 \quad \left(\text{Enstad's equation 16, [ENS 75]}\right)$$

The unconfined stress f_c varies as a function of σ_1 according to the approximate law (valid for $\sigma_1 > \sigma_{1m}$):

$$f_c = f_{co} + k\sigma_1$$

$$\sigma_1 = (1+\sin\delta)\bar{\sigma}(r)$$

For that to happen, we simply have to replace β_p with α and $\bar{\sigma}$ with $(\sigma_1 + \sigma_2)/2$. We then have the following overall equation:

$$\Delta W + \Delta F_2 = \Delta F_1 \quad \left(\text{Enstad's equations C4 and C5, [ENS 75]}\right)$$

We then eliminate σ_1 by:

$$\sigma_1 = f_{co} + k\left(1 + \sin\delta\right)\left(\frac{\sigma_1 + \sigma_2}{2}\right) + \frac{1 + \sin\phi}{1 - \sin\phi}\sigma_2$$

This results in the differential equation:

$$r\frac{d\sigma_2}{dr} - X_b\sigma_2 = -\gamma Y_b r + Z_b r^X + Z_o$$

The solution of this differential equation is [ENS 75, Appendix E]:

$$\sigma_2(r) = a\left[\frac{r}{R}\right]^X + b\left[\frac{r}{R}\right]^{X_b} + c\left[\frac{r}{R}\right] + d$$

According to Enstad [ENS 75, Appendix F], we just have to keep the last two terms since X and X_b are of the order of 10:

$$\sigma_2(r) = c\left[\frac{1}{R}\right] + d$$

where:

$$c = \frac{\gamma Y_b R}{X_b - 1} \qquad d = -\frac{Z_o}{X_b} \qquad u = 1 + \frac{\sin\left(2\alpha + \theta\right)}{\sin\theta}$$

Rectangular cross-section   Circular cross-section

(Enstad's equation 23 and 24, [ENS 75]) [ENS 77], equation 21 and 22

$$X_b = \frac{u\sin\phi}{1 - \sin\phi} \qquad\qquad X_b = \frac{2u\sin\phi}{1 - \sin\phi}$$

$$Y_b = \frac{(\alpha+\theta)\sin\theta}{\sin^2(\alpha+\theta)} + \frac{\sin\alpha}{\sin(\alpha+\theta)} \qquad Y_b = \frac{\sin\alpha + \sin\theta\left[1+\tan^2\left(\dfrac{\alpha+\theta}{2}\right)\right]}{\sin(\alpha+\theta)}$$

$$-\frac{ku\,Y}{2(X-1)}(1+\sin\delta) \qquad\qquad -\frac{ku\,Y}{(X-1)}(1+\sin\delta)$$

$$Z_o = \frac{uf_{co}}{2} \qquad\qquad\qquad Z_o = uf_{co}$$

The equations for X and Y have been given regarding the determination of stresses (see Chapter 2).

If an arch exists, the lower surface does not support a normal stress, that is, σ_2 is zero. We can deduce the minimal distance r_m from the tip of the convergent to the outlet:

$$r_m = -\frac{Rd}{c} = \frac{(X_b-1)Z_o}{X_b\gamma Y_b}$$

According to Enstad, Jenike's results correspond to:

$$r_m' = \frac{Z_o}{\gamma Y_b}$$

This modification is aimed at increasing security since it leads to an increase in r_m. Finally:

Rectangular cross-section   Circular cross-section

Width : $B = \ell_m = 2r_m\sin\theta$ Diameter : $B = D_m = 2r_m\sin\theta$

More generally:

$$r_m = \frac{B}{2\sin\theta}$$

NOTE.–

We can assume that:

$$\sigma_1(r) = (1 + \sin\delta)\,\gamma\,r\,s'(r)$$

But, according to Enstad:

$$\sigma_1(r) = (1 + \sin\delta)\,\vec{\sigma}(r) = (1 + \sin\delta)\,\gamma r\left[\frac{Y}{X-1}\right]$$

The expression of s' is therefore:

$$s' = \frac{Y}{X-1}$$

EXAMPLE 3.6.–

Calculation of the minimum diameter of the withdrawal opening.

$\theta = 15°$ $\delta = 45°$ $\phi = 20°$ $\phi_p = 15°$

$R = 7.7279$ m $\gamma = 11{,}772$ N.m^{-3} $f_{co} = 4{,}724$ Pa $k = 0.15$

$X = 19.422$ $Y = 3.7049$ $\sin\theta = 0.2588$

The parameters which come into picture are:

Since : $\alpha_M = 45° - 7.5° = 37.5°$ and $37.5° > \phi_p = 15°$

$\alpha = 15°$

$$u = 1 + \frac{\sin(30° + 15°)}{\sin 15°} = 3.7320$$

$$X_b = \frac{2 \times 3.732 \sin 20°}{1 - \sin 20°} = 3.8796$$

$$Y_b = \frac{\sin 15° + \sin 15°(1 + \mathrm{tg}^2 15°)}{\sin 30°} - \frac{0.15 \times 3.732 \times 3.7049(1 + \sin 45°)}{19.422 - 1}$$

$Y_b = 0.88025$

$Z_o = 3.732 \times 4\,724 = 17\,629$

$c = (11,\,772 \times 0.88025 \times 7.7279/2.8796 = 27,\,809$

$d = -17,\,629/3.8796 = 4\,544$

$$r_m = \frac{7.7279 \times 4544}{27.809} = 1.26270 \text{ m}$$

$$r_m' = \frac{17\,629}{11.772 \times 0.88025} = 1.70126 \text{ m}$$

Hence the minimum diameters for which arching does not occur are:

According to Enstad: $D_m = 2 \times 1.26270 \times 0.2588 = 0.6535 \text{ m}$

According to Jenike: $D_m' = 2 \times 1.70126 \times 0.2588 = 0.8806 \text{ m}$

3.5.2. Graphical procedure [JOH 64b]

Let us find the minimum range of an arch. If dy represents the vertical thickness of the arch, its weight is:

$\gamma \, Ady$

A: area of the horizontal projection of the arch (m^2)

γ: specific weight of the product ($N.m^{-3}$)

The thrust of vertical reaction of the support is, at most:

$$f_c \left[dy \, \cos\left(\frac{\pi}{2} - \alpha \right) \right] P \, \cos\alpha = \frac{f_c P \, \sin 2\alpha dy}{2}$$

f_c: unconfined yield stress (Pa)

P: perimeter of the arch (m)

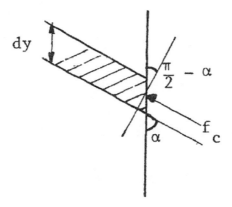

Figure 3.10. *Support of an arch*

Let us say:

$$\frac{A}{P} = \frac{B}{2(1+m)}$$

m = 1 for a circular cross-section for which B is the diameter

m = 0 for a slot for which B is the width.

Let us now equal the weight with the thrust of the reaction:

$$B = \frac{(1+m)f_c\sin 2\alpha}{\gamma}$$

The upper limit of sin2α is 1, which corresponds to α = 45°. We will consider this value such that, in order for the arch to collapse:

$$B > \frac{(m+1)f_c}{\gamma}$$

But, we know that the major principal stress σ_1 is:

$$\sigma_1 = \left(1 + \sin\delta\right)\gamma r_m s'\left(r_m\right) = \frac{\left(1 + \sin\delta\right)\gamma B s'\left(r\right)}{2\ \sin\theta}$$

Hence:

$$B = \frac{2\ \sin\theta\sigma_1}{\left(1 + \sin\delta\right)\gamma s'} > \frac{\left(m + 1\right)f_c}{\gamma}$$

That is:

$$\frac{\sigma_1}{f_c} > \frac{\left(1 + \sin\delta\right)\left(m + 1\right)s'}{2\ \sin\theta} = k_{Emin}$$

Note that this inequality defines a minimum value of the flowability ratio of the product (see section 1.3.7):

$$\frac{\sigma_1}{f_c} = k_{E\ prod} > k_{Emin}$$

If we know the function $f_c\left(\sigma_1\right)$, we can plot the following graph:

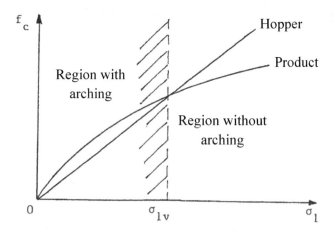

Figure 3.11. *Graphical condition for non-arching*

Thus, we see that:

Arching occurs if $\sigma_1 < \sigma_{1v}$

Flowing is possible if $\sigma_1 > \sigma_{1v}$

NOTE.– Simplified calculation of the diameter of the minimum opening diameter:

If we know, fairly precisely, the unconfined yield stress f_c, Jenike suggests that we use:

$$D_m = \frac{2.2\ f_c}{\gamma} \quad (\text{for a circular opening})$$

3.5.3. Slenderness of a silo and arching

A silo is said to be slender when the ratio of its height H to its diameter is high. More precisely:

$$H > 15(A / P)$$

A and P are respectively the area and perimeter of the horizontal cross-section of the silo or hopper.

Note that, in these conditions, the exponential which appears in Janssen's formula can be neglected and the average stress $\overline{\sigma}$ is then at its maximum value (see sections 2.1.3 and 2.1.4).

$$\overline{\sigma} = \frac{\gamma A}{P\,\mathrm{tg}\theta_p\left(1 + \sin\theta\,\cos2\beta\right) - \sigma_o}$$

If, on the other hand, the vertical wall is smooth, that is if $\mathrm{tg}\phi_p$ is small, that would be a second reason for $\overline{\sigma}$ to be high.

Thus, the product arrives highly consolidated at the cylinder–convergent boundary (its f_c is high). At the crossing of this boundary, arching can occur.

The solution is to increase the friction in the vertical portion,

– either by increasing the roughness of the wall;

– or by introducing horizontal iron bars across the silo.

In fact, we know that, in the passive state which will have become that of the cylinder, the major principal stress σ_1 is equal to σ_h:

$$\sigma_1 = \sigma_h = \frac{A\gamma}{P\tan\phi_p}\left[1 - \exp\left(-\frac{PBDh}{A}\right)\right]$$

If ϕ_p increases, σ_1 decreases as well as the consolidation and consequently f_c. The flowability $C = 1/f_c$ is therefore improved.

3.5.4. Stresses in a downward convergent

Jenike [JEN 64a, JEN 64b] defined the function $s(\theta)$ as follows:

$$\bar{\sigma} = \frac{\sigma_1 + \sigma_2}{2} = \gamma rs(\theta)$$

We see that the stresses are proportional to the polar radius r.

σ_1 and σ_2 are the principal stresses in two dimensions or in a meridian plane in three dimensions.

γ: specific gravity of the D.S. $(N.m^{-3})$

θ: polar angle between the vertical axis and a radius r

Jenike [JEN 64a, JEN 64b, JEN 65] specified the calculation of $s(\theta)$ by using the mathematical method of characteristics. In his Figures 8, 9, and 10 of [JEN 64a], he gives the representation of the variations of $s(\theta)$ in terms of θ and of the friction angle ϕ of the divided solid on the wall of the convergent and that for a given value of the dynamic friction angle δ of the D.S. which is assume to be equal to $50°$.

– Figure 8 in two dimensions;

– Figure 10 in two dimensions but with a vertical wall;

– Figure 9 representing a rotation symmetrical convergent in three dimensions.

3.5.5. Characteristic ratios and condition for flow without interruption

Johanson and Colijn [JOH 64b] started by defining the consolidation pressure σ_1 around the outlet opening

$$\sigma_1 = \frac{\gamma B s_p \left(1 + \sin\delta\right)}{2\sin\theta_p} \quad \text{with} \quad s_p = s\left(\theta_p\right)$$

s_p: value of s on the wall

θ_p: angle made by the wall and the vertical axis

B: size of the opening: width in 2D and diameter in 3D (m)

The characteristic ratio of the convergent, referred to as ff by the authors, is given by:

$$ff = \frac{\sigma_1 \left(1 + m\right)}{\gamma B}$$

– two dimensional convergent: m = 0; B is the width of the opening;

– conical rotation convergent: m = 1; B is the diameter of the opening.

The ratio ff can be read from Johanson and Colijn's Figures 6, 7, 8, and 9 [JOH 64b] in terms of the internal friction angle ϕ and the dynamic friction angle (see section 1.3.11 and Figure 1.14) (the authors call it "efficient") of the D.S. The figures correspond to the following values of the angle of slope θ_p of the wall with the vertical axis respectively:

$$\theta_p = 0, 10, 20, 30 \text{ sexagesimal degrees}$$

The ratio of flowability for the D.S. is:

$$k_E = \frac{\sigma_1}{f_c(\sigma_1)}$$

The flow through the opening will be possible if the D.S. yields to the influence of gravity.

$$k_E > ff$$

In other words, the condition for flow without interruption is:

$$f_c < \frac{\gamma B}{(1+m)}$$

f_c: unconfined stress of the D.S. (Pa) (see section 1.3.6 and Figure 1.10 in [DUR 16])

In other words, the effect of the specific gravity multiplied by the size of the opening must be greater than the unconfined stress f_c by the factor 1/(1 + m).

NOTE.–

We have seen that Enstad shows that Jenike and Johanson's method described above leads to oversizing of the opening. He therefore suggests a more accurate method for calculating the size of the opening (see section 3.5.1):

3.6. Rate of emptying

The rate of emptying of a silo depends only on the state of the divided solid around the outlet. In fact, immediately after opening, flowing begins and a strong dilatation or expansion wave spreads upwards up to the free surface.

3.6.1. *Rate of mass drainage*

Hagen [HAG 52] puts forth the classical expression of this rate. It was not until 1961 that Brown proved this expression theoretically by establishing that it corresponds to a minimum value of the total Helmatlz Hel energy of the D.S. (sum of its potential energy and its kinetic energy). The rate is proportional to:

Circular opening of diameter D: $\qquad \rho_a \sqrt{g}\left(D - kd_p\right)^{5/2}$

Slot opening of width ℓ and length L: $\quad \rho_a \sqrt{gL}\left(1 - kd_p\right)^{7/2}$

Beverloo *et al.* [BEV 61] confirmed the expression relating to the circular opening.

ρ_a: apparent mass density of the D.S. in bulk (kg.m^{-3})

g: acceleration due to gravity: 9.81 m.s^{-2}

d_p: average diameter of particles (m)

$$d_p = \Sigma m_i d_{pi}$$

m_i: mass fraction of the particles of diameter d_i

Savage and Sayed [SAV 81] defined the mass rate of emptying across a circular hole.

$$W = J\frac{\pi}{4}\rho\sqrt{g}\left(D - kd_p\right)^{5/2}$$

These authors give, alternatively, two complex expressions for J:

J_1 given by their equation 33

J_2 given by their equation 43

J_1 and J_2 depend on the following angles:

Internal friction of the D.S.

Of friction on the wall

Of the wall with the vertical axis

Angle of approach (between the vertical axis and the moving-stationary boundary in the D.S.). Between the major principal stress and the radial direction making an angle θ with the vertical axis.

Using the results obtained by Shinohara *et al.* [SHI 68], by improving or completing and specifying them, Firewicz [FIR 88] suggests his equation (33) for the rate of mass emptying.

$$W = 0.453 \rho_a \sqrt{g} \left(tg\theta \right)^{-0.5} D_o^{2.5}$$

On the other hand, Firewicz [FIR 88] gives the value of the average speed U_A of the particles at the level of the dome which is formed periodically above the outlet.

$$U_A = 0.325 \sqrt{g} \left(\frac{D_o}{k} \right)^{0.5} \left(\frac{1-\varepsilon_o}{1-\varepsilon_A} \right)$$

He also gives the expression for the pulse rate due to the formation and destruction of arches or domes.

$$n = \frac{1}{\ell\sqrt{3}} \sqrt{\frac{gD_o}{k}} \left(\frac{1-\varepsilon_o}{1-\varepsilon_A} \right)$$

Firewicz [FIR 88] does not say it but acknowledges in his calculation of the frequency n that $\ell = D_o$ and he finds a value in the order of 10 Hz.

ε_o: porosity at the opening (bulk porosity)

ε_A: porosity at the flattened dome (compact porosity)

ℓ: height of the dome: m

D_o: diameter of circular opening (m)

g: acceleration due to gravity: 9.81 m.s^{-2}

k = tanθ

θ: angle between the vertical axis and wall of the cone

3.6.2. *Johanson's expression [JOH 65] for the rate of emptying*

By examining the acceleration of particles when an arch or dome collapses above the outlet, the author obtains the following expression for the rate of instantaneous mass emptying across the outlet:

– the outlet is a slot of length L and width B

$$W = \gamma BL \sqrt{\frac{Ba}{2\,tg\,\theta_p}} \qquad\qquad \left(kg.s^{-1}\right)$$

– the outlet is a circle of diameter B:

$$W = \gamma \frac{\pi B^2}{4} \sqrt{\frac{Ba}{4\,tg\,\theta_p}} \qquad\qquad \left(kg.s^{-1}\right)$$

θ_p: angle between the conical wall with the vertical axis

γ: specific gravity of the D.S. $(N.m^{-3})$

According to the author, the average rate in time is not very different from the instantaneous rate. On the other hand, we will be able to liken, apparently, the acceleration a to that of gravity g.

The author derives the expression for the unconfined stress f_c of the D.S. with the help of his equations 22 and 23, from the measure of the flow rate W and compares it with the values obtained from Mohr's circle derived with the help of a shear cell. We will refer to the original publication, which shows that there is a match.

NOTE.–

Firewicz [FIR 86] accounts for the effect of the diameter T of the capacity with flat bottom.

$$W = \sqrt{g}\rho_a d_p^{2.5} \times i \left(\frac{D_o}{d_p}\right)^j$$

$$i = 0.811T + 0.101$$

$$j = 2.98 - 1.69T$$

NOTE.–

Dehne and Keller [DEH 85] give an empirical expression for the flow rate in terms of the geometric dimensions of the hopper. Zenz [ZEN 57a, ZEN 57b] had also proposed an expression.

NOTE.–

Savage [SAV 65], with the help of a simplified theory, suggested the following expression for the mass flow rate:

$$W = \frac{\pi}{4}\rho_a D^{2.5}\sqrt{g}\left[\frac{(1+k)}{2(2k-3)\sin\theta_p}\right]^{1/2}$$

with:

$$k = \frac{1+\sin\phi}{1-\sin\phi}$$

ϕ: internal friction angle

θ_p: angle between the vertical axis and the wall

3.6.3. Effect of the product on the parameters of mass desilage flow rate

1) The shape of the particles has little or no effect of the flow rate.

2) The surface roughness of the particles increases the internal friction angle ϕ.

3) The angularity (presence of sharp edges and tips) increases the porosity (volume void fraction).

4) The dynamic friction angle δ increases if the size of particles reduces, because the number of mutual contact points, that is friction points, increases. An amount of 20% of fly ash at 20 μm in the sand at 850 μm sharply reduces the desilage flow rate.

3.6.4. Effect of the diameter of particles on the flow rate

Recall that the weight of a particle of diameter d_p is:

$$\pi_p = \frac{\pi}{6} d_p^3 \rho_s g$$

Rose and Tanaka [ROS 59] make use of a coefficient c, which they call cohesion but which, in reality, is a force. They do not give the expression of c. But for that we can call upon Hamaker's force F_H ([HAM 37]; see 5.3.1 volume 7).

$$F_H = \frac{A d_p}{24 e^2}$$

A: Hamaker's constant: $0.7.10^{-19}$ J

e: distance between the surfaces of two neighboring particles: e = 50 nm

So:

$$F_H = 1.17 \, d_p$$

Hence the following expression:

$$\frac{\pi F_H}{\pi_p} = \frac{1.17 d_p \times 6}{g \, \rho_s d_p^3} = \frac{0.715}{\rho_s d_p^2}$$

Rose and Tanaka's [ROS 59] coefficient 7.7×10^{-6} cannot be conserved and we will propose the coefficient as the multiplication factor ϕ_d of rate of drainage.

$$\phi_d = \exp\left(-\frac{1.9 \times 10^{-9}}{\rho_s d_p^2} \right)$$

EXAMPLE 3.7.–

$$\rho_s = 2.700 \text{ kg.m}^{-3} \qquad d_p = 10^{-6} \text{ m}$$

$$\phi_d = \exp\left(-\frac{1,9.10^{-9}}{2700 \times 10^{-12}}\right) = \exp(-0.7)$$

$$\phi_d = 0.49$$

Due to the lack of accurate tests, *the value 1.9×10^{-19} is just an order of magnitude*.

3.6.5. Small-sized circular opening

In order to avoid obstacles due to cross-bridging, the minimum diameter of the opening must be greater than $6d_p$ where d_p is the diameter of the largest particles.

This non bridging criterion is much less acute than the nonarching criterion (which leads to much larger openings). The non bridging criterion only applies to highly flowing products (granules).

For an opening which meets this criterion, we see that a region is neutralized at the edge of the opening, and according to Firewicz's [FIR 84] equation 17 we can write:

$$D'_o = D_o - 1.56\frac{d_p}{\phi_s}$$

ϕ_s is the ratio of sphericity:

$$\phi_s = \frac{\text{surface area of the sphere with equal volume}}{\text{surface area of the particle}^{(1)}}$$

(1) measured by filtration in a laminar or by gas adsorption.

3.6.6. Effect of the roughness of the wall of the convergent on the flow rate

A rough wall obviously slows the product down (but only in its immediate surrounding) and can even lead to the formation of a stagnant layer, which, on the contrary, increases the overall flow rate, since the presence of this dead zone causes the half-angle at the top of the convergent θ to assume an effective value of θ_e, which is smaller, as we can see from Figure 3.12. In fact, we know that the flow rate is directly proportional to $(tg\ \theta)^{-1/2}$.

If on the periphery of the outlet there is a region not covered with stationary product, we notice that the extension of this region increases with the smooth nature of the wall. Conversely, the fraction of the wall covered with a dead zone increases as the roughness increases since, at the same time, θ_e reduces. Note that the covered portion also increases with the value of θ_e.

Naturally, if the wall lies entirely below the dead zone, the changes in the parietal roughness no longer have an effect on the flow rate. This is the case with great angle convergents ($\theta < 85°$) or flat-bottom convergents.

Nguyen *et al.* [Ngu 80] distinguish five types of flow and state their existence in the plane with θ on the abscissa and the ratio H/W on the ordinate axis.

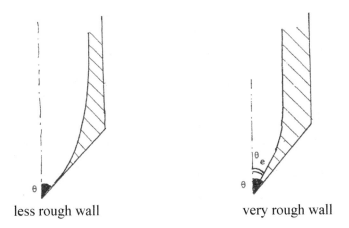

less rough wall   very rough wall

Figure 3.12. *Flow with a rough wall on the convergent*

3.6.7. Effect of air pressure

Below are three publications discussing this effect:

– Resnick *et al.* [RES 66];

– Sutton and Richmond [SUT 73];

– Nedderman *et al.* [NED 83].

3.7. Withdrawal of fine products

3.7.1. Mass flow mechanism of fine products

Under hoppers of mass flow, we generally provide an extractor directly under the opening, which normally discharges the flow rate of the solid, which is less than the maximum flow rate corresponding to the size of the opening. If we increase the speed of the extractor, the flow rate increases until a certain limiting value is reached, an arch of product is formed above the opening, and a particle shower comes down from it while the arch is stabilized by the air friction, which acts against the particle shower. The flow under this arch is also slowed down by the air.

In fact, air must flow into the convergent in order to occupy the volume left empty by the withdrawn solid.

Naturally, the friction between the gas and the solid is especially high as the latter is less permeable as particles become finer.

The products exhibiting this behavior are, for example, kaolin, fine ground feldspar, activated earth, and, in general, the products whose average harmonic diameter of the particles is less than 80 μm.

3.7.2. Permeability, specific resistance, and law of percolation

The permeability K is defined by [DUR 99]:

$$K = \frac{\varepsilon^3 d_p^2}{150(1-\varepsilon)^2}$$

K: permeability (m^2)

d_p: average harmonic diameter of particles (m)

ε: porosity of solid medium.

The specific resistance α is defined by

$$\alpha = \frac{150(1-\varepsilon)}{\rho_s d_p^2 \varepsilon^3} = \frac{1}{K\rho_s(1-\varepsilon)}$$

α: specific resistance ($m.kg^{-1}$)

ρ_s: actual mass density of the solid ($kg.m^{-3}$)

The law of percolation (Darcy) is then written for a fluid, in this case a gas as:

$$\frac{dP}{dy} = \frac{\mu V_G}{K} = \mu V_G \alpha \rho_s (1-\varepsilon) \qquad [3.2]$$

V_G: speed of gas in empty container ($m.s^{-1}$)

μ: viscosity of the gas (Pa.s)

g: acceleration due to gravity: $9.81\ m.s^{-2}$

The friction ΔP is equivalent to a fictitious drop Δg of the gravitational field:

$$\frac{dP}{dy} = \Delta g \rho_s (1-\varepsilon) \qquad [3.3]$$

Hence, equating [3.2] and [3.3]:

$$\Delta g = \mu \alpha V_G$$

The effective gravitational field g_e is:

$$g_e = g - \mu \alpha V_G$$

3.7.3. Compactness of confined dispersion in free fall

Let us consider 1 m^3 of dispersion of solid particles in free fall.

We will take into consideration the fact that the thermodynamic law of entropy maximization is applicable for this volume. In other words, the compactness β of the suspension (1's complement of the porosity ε) is such that the energy loss is maximum. This energy loss corresponds to the drop in total free energy.

During the time τ, the vertical distance covered by the dispersion cloud is $\tau V_{\ell N}$ and the gravitational potential energy dissipated is:

$$\beta \tau V_{\ell N} g \rho_S$$

But, the speed limit $V_{\ell N}$ of the fall of particles in the cloud is:

$$V_{\ell N} = V_\ell (1 - \beta)^n$$

V_ℓ: speed limit of the free fall of an isolated particle (m.s^{-1})

β: compactness of the particle cloud (fraction of solid in the volume)

We therefore have to maximize the function:

$$(\tau V_\ell g \rho_S) \left[\beta (1 - \beta)^n \right]$$

The maximum value is obtained for:

$$\beta = \frac{1}{n-1}$$

For fine particles, Richardson and Zaki's correlation gives:

$$n = 4.4 \qquad \beta = 0.294 \qquad \varepsilon = 0.706$$

The apparent mass density ρ_a is therefore:

$$\rho_a = \rho_S (1 - \varepsilon) = \rho_S \beta$$

ρ_S = actual density of the product (kg.m^{-3})

ε = porosity

However, at the opening, the porosity has been measured by Delaplaine [DEL 56] who, on the other hand, suggests that we adopt the value ε_{fc} of the starting fluidization. In these conditions, the weight of the solid is compensated by the drop in the gas pressure:

$$g\left(1-\varepsilon_{fc}\right)\rho_S = \frac{\Delta P}{\Delta y}$$

The porosity ε_{fc} increases if the sphericity decreases and the angularity increases. The same applies to the porosity at rest (in bulk) ε_0.

Nature of the solid	ε_0
Spheroidal sand	0.4
Angular sand (ground)	0.5

Table 3.1. *Bulk porosity*

3.7.4. *Flow rate of solid retarded by the rise of air*

The gas flow rate must compensate for the volume left by the withdrawn solid:

$$A_o\left(\frac{V_G}{\varepsilon} - V_S\right) = A_o V_S\left(1-\varepsilon\right)$$

Hence:

$$V_S = V_G$$

V_S: speed of solid particles (m.s^{-1})

V_G: speed of gas in empty container (m.s^{-1})

The acceleration due to gravity g is reduced by the gas friction and becomes (see section 1.7.2 in [DUR 99]):

$$g\left(1 - \frac{\alpha\mu V_G}{g}\right) = g(1 - X) \quad \text{with} \quad X = \frac{\alpha\mu V_G}{g}$$

α: specific resistance of the solid $(m.kg^{-1})$

μ: viscosity of the gas (Pa.s)

The rate of flow of the solid during a fall is directly proportional to the square root of the acceleration due to gravity. Therefore:

$$V_S = V_{so}(1 - X)^{1/2} \quad \text{hence}: \frac{\alpha\mu V_G}{g} = \frac{\alpha\mu V_S}{g} = \frac{\alpha\mu V_{So}}{g}(1 - X)^{1/2}$$

Let us say:

$$k = \frac{\alpha\mu V_{So}}{g}$$

Therefore:

$$X^2 = k^2(1 - X)$$

$$X = \frac{k^2}{2}\left[\left(1 + \frac{4}{k^2}\right)^{1/2} - 1\right]$$

The gas flowrate is therefore:

$$V_G = \frac{gX}{\alpha\mu}$$

3.7.5. Drop in gas pressure above the opening

The law of percolation cane be written as:

$$\frac{dP}{dr} = \frac{\mu}{K}V_G \quad \text{with}: V_G = \frac{W_G}{\rho_G A_o}$$

P: pressure (Pa)

r: distance at the top (fictitious) of the convergent (m)

μ: viscosity of the gas (Pa.s)

K: permeability of the solid above the opening (m^2)

V_G: speed of the gas: in an empty container ($m.s^{-1}$)

W_G: massive gas flow rate ($kg.s^{-1}$)

A_o: cross-sectional area of the opening (m^2)

ρ_G: density of the gas ($kg.m^{-3}$)

Eliminating V_G:

$$\frac{dP}{dr} = \frac{\mu W_G}{K\rho_G A_o} \qquad A_o = \pi r^2 tg^2\theta$$

θ: half-angle at the top of the convergent:

$$\frac{dP}{dr} = \left[\frac{\mu W_G}{K\rho_G \pi tg^2\theta}\right]\frac{1}{r^2}$$

Let r_o be the radial spacing of the opening. Integrating:

$$P - P_G = \left[\frac{1}{r_o} - \frac{1}{r}\right]\left[\frac{\mu W_G}{K\pi\rho_G tg^2\theta}\right]$$

We can take into consideration the fact that the zone in which the product collapses exists between the following radial spaces:

$$r_o = \frac{B_o}{2tg\theta} \quad \text{and} \quad r = r_o + B_o$$

EXAMPLE 3.8.–

$\rho_G = 1.2$ kg.m^{-3} $\theta = 9.68$ degr.

$d_p = 285$ μm $= 0.285.10^{-3}$ m

$D_o = 0.01$ m $\mu = 18.10^{-6}$ Pa.s

$\rho_S = 2.570$ kg.m^{-3} $\varepsilon = 0.70$

$$A_o = \frac{\pi}{4} \times (0.01)^2 = 0.785.10^{-4} \text{m}$$

$$K = \frac{(0.7)^3 (0.285.10^{-3})^2}{150(1-0.7)^2} = 2.0637.10^{-9} \text{m}^2$$

$$\alpha = \frac{1}{2.0637.10^{-9} \times 2570(1-0.7)} = 6.2849.10^5$$

$$\rho_a = 2570(1-0.7) = 771 \text{ kg.m}^{-3}$$

From section 3.6.2:

$$V_{S0} = \left[\frac{0.01 \times 9.81}{4 \times 0.1705}\right]^{1/2} = 0.379 \text{ m.s}^{-1}$$

$$k = \frac{6.2849.10^5 \times 18.10^{-6} \times 0.379}{9.81} = 0.43706$$

From section 3.7.4:

$$X = 0.3518$$

$$V_G = \frac{9.81 \times 0.3518}{6.2849.10^5 \times 18.10^{-6}} = 0.305 \text{ m.s}^{-1}$$

$$W_S = 2570(1-0.7)0.785.10^{-4} \times 0.379(1-0.3518)^{1/2}$$

$$W_s = 0.01846 \text{ kg.s}^{-1} \quad r_o = \frac{0.01}{2 \text{ tg} 9.68°} = 0.028 \text{ m}$$

$$r = r_o + D_o = 0.038 \text{ m}$$

$$\frac{18.10^{-6} \times 0.305 \times 1.2 \times 0.785.10^{-4}}{2.0637.10^{-9} \times \pi \times 1.2 \times 0.0291} = 2.2843$$

$$P - P_o = 2.2843 \left[\frac{1}{0.028} - \frac{1}{0.038} \right] = 82 - 60 = 22 \text{ Pa}$$

3.7.6. Powders and their permeability to gases

Let us assume that the gaseous sky of the silo is isolated from the atmosphere because it consists of an inert gas. During the emptying, some gas will have to get inside the silo through the outlet to take the place of the D.S., which will have been removed. This slows down the disposal of the solid.

Ducker *et al.* [DUC 85] came up with the following equation for the slower flowrate:

$$W = W_o \left[\frac{\sqrt{4 + E_1^2 W_o^2} - E_1 W_o}{2} \right]$$

This equation is derived from similar assumptions made to those of section 3.7.2.

With:

$$E_1 = \frac{150 c_1 \mu}{2 \pi r_o^2 \rho_s^2 g \varepsilon^3 d_p^2 (1 - \cos\theta)} ; \quad c_1 = \frac{2K - 3}{2K - 1} ; \quad K = \frac{1 + \sin\phi}{1 - \sin\phi}$$

θ: half-angle at the top of the lower cone of the silo

ϕ: internal friction angle of the D.S.

r_o: distance between the tip (virtual) of the cone and the edge of the outlet (m)

$\overline{d}_p$: average harmonic diameter of particles (m)

$$\frac{1}{\overline{d}_p} = \sum_i \frac{m_i}{d_{pi}}$$

m_i: mass fraction of the particles of diameter d_{pi}

ε: porosity of the D.S. in the hopper

g: acceleration due to gravity: 9.81 m.s^{-2}

W_o: mass flowrate if the silo is not isolated from the atmosphere

W_o can be calculated using one of the equations from section 3.6.1.

Johanson [JOH 79a, JOH 79b] reviews the cases where a gas–solid flow occurs and suggests remedies in some of the cases.

3.7.7. Flow of fine products with the presence of a well

During withdrawal, the upper portion of the well momentarily and rapidly collapses taking an influx of the product to the extractor, which could block the latter. In fact, during the collapse, the product still causes air to flow and the mass is left very aerated. It can therefore flow like a liquid. This results in a momentary overflow followed by a complete interruption. In these conditions, the right setting of the flow rate becomes difficult if not impossible, even if we interpose a rotary valve in between the silo and the extractor.

However, in mass flow hoppers, powders do not show this phenomenon if the storage time was large enough to enable the degassing of interstitial air and the discharge then takes the form of an extrusion.

The solution is therefore to find the mass flow.

3.8. The kinematic theory of flow in a hopper

3.8.1. Flow in a vertical cylinder

Let us consider a divided solid made up of sites which could be empty or occupied by a particle. The number of sites per m^3 of D.S. is given by:

$$n_s = \frac{1}{\omega_s} = c_v + c_p \quad \text{with} \quad c_v \ll n_s$$

ω_s: volume of a site: m^3

c_v and c_p: concentration of empty sites and concentration of occupied sites (m^{-3})

During the gravity flow of a D.S. in a vertical cylinder, the concentration c_v of empty sites satisfies the following equation [MUL 72].

$$D\frac{\partial^2 c_v}{\partial x^2} + D\frac{\partial^2 c_v}{\partial y^2} = V\frac{\partial c}{\partial z}$$

Graham *et al.* [GRA 87] give the solution to this equation with:

D: diffusivity $(m^2.s^{-1})$

V: speed $(m.s^{-1})$

They derive an equation for the vertical speed at every point of a horizontal section for smooth particles as well as for rough particles.

Tüzün and Nedderman [TUZ 82b] confirmed the validity of the kinematic theory for granules.

3.8.2. Flow in a vertical convergent

Tüzün and Nedderman [TUZ 79] establish that the equation of the problem is parabolic for the vertical speed v.

$$B\frac{\partial^2 v}{\partial x^2} = \frac{\partial v}{\partial y} \quad \text{with} \quad B = kd_p$$

y: vertical coordinates: m

x: horizontal coordinates: m

B: kinematic constant: m

d_p: diameter of particles: m

The coefficient k depends on the shape of the particles.

For particles for which the diameter is greater than 500 μm, we can assume that the interstitial air has no effect on the flow.

The local velocity is represented by a bell-shaped curve as a function of the distance to the axis. The equation of this curve is given by the equation 5 of [TUZ 79].

3.8.3. Advantages of the kinematic theory

Tüzün *et al.* [TUZ 82a] show that this theory

– enables us to predict the velocity profile in a vertical convergent;

– enables us to predict the extent of the stagnant zone if it exists;

– makes use of much simpler equations than the theory of plasticity does.

An example of the application of the theory of plasticity is given by Nguyen *et al.* [NGU 79].

NOTE.–

Davidson and Nedderman's [NED 73] theory of the watchglass suggests an equation for the rate of withdrawal but this equation is valid for the assumption that:

– the walls of the hopper are smooth;

– the half-angle at the top of the convergent is small; and

– the D.S. is without internal cohesion.

3.9. Activation of the emptying

3.9.1. *Activation by vibrations*

We can consider the following solutions:

1) Installation of the hopper assembly on flexible supports with flexible links for the inlet and outlet. An eccentric system produces vibrations.

$$\text{frequency} < 50\,\text{Hz} \quad \text{and} \quad \text{amplitude}\,A < 8\,\text{mm}$$

This system allows for large silage heights even for products which, during storage, would be more cohesive and consolidate into blocks or clods.

However, the vibrations can "overconsolidate" certain products, that makes them compact and increases their cohesion instead of reducing it. This leads to the formation of wells and a reduction in the flowrate, which can become erratic.

2) In order to remedy this drawback, we can limit the vibration to the lower portion of the hopper.

3) Installation of perforated plates or expanded metal slabs vibrating parallel to the wall and immersed in the product. This device is very silent.

3.9.2. *Flow promoter slabs (vibrated or nonvibrated)*

We know that the minimum width of a hopper's outlet opening is inversely proportional to the acceleration due to gravity.

So, if we impart vertical vibratory motion to the base of a hopper, the acceleration due to this motion will add to the acceleration due to gravity and cause the collapse of the arch. The acceleration due to the vibrations is of the form:

$$a = -A\omega^2 \sin\omega t$$

Apparently it is easier to create horizontal vibrations. That is why we adopt the cone-like slab [JOH 66a, JOH 66b, WAH 81, TÜZ 83].

A horizontal acceleration can only cause horizontal and not vertical motion. But the oblique flow around the cone has a vertical component.

Besides, the presence of the slab cancels the static compression effect above the outlet, this compression being the cause of arching by an increase of the rigidity domain of the product. In fact, the vertical friction is increased by the presence of the cone, which similarly reduces the effect of compression due to the weight of the product.

Figure 3.13. *Flow shield agent*

The vibrators can cause setbacks and make the product more compact instead of breaking the arch if the flow rate of the discharge carrier is insufficient and does not "follow" the product's descent.

It is also possible to avoid arching by inserting nonvibrated shields. Johanson's publication [JOH 66a, JOH 66b] gives indications of the dimensioning and optimal position of these flow improving agents. In brief:

– If we want the product to go down the cylinder in a block, a large cone need to be installed, close to the cylinder-convergent transition point and below the latter.

– If we want to avoid the occurrence of arching and/or pit flow, we should then provide a small diameter cone close to the outlet.

3.9.3. *Air injection and angle at the tip of the convergent*

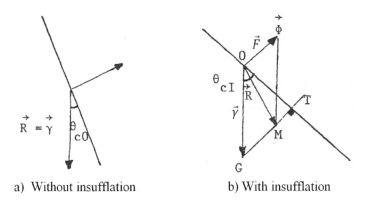

a) Without insufflation   b) With insufflation

Figure 3.14. *Consequence of insufflation*

In order for the sliding on the wall to keep up, it is necessary for the resultant $\vec{R} = \overline{OM}$ of the volume forces exerted on the solid to make an angle less than $\dfrac{\pi}{2} - \phi_p$ with the wall, where ϕ_p is the friction angle.

$\vec{R} = \overline{OM}$ is the resultant of the specific gravity γ of the product and of the volume force F due to friction of the injected gas across the product.

Without insufflation, the resultant $\vec{\gamma}$ is vertical. The half-angle at the top of the convergent θ_{cO} must therefore satisfy the condition:

$$\theta_{c0} < \frac{\pi}{2} - \phi_p$$

If, on the contrary, gas (air) is injected in the hopper across porous panels consisting of the wall of the convergent of the latter, the resultant $\vec{R}$ will no longer be vertical, as we see in Figure 3.14(b), and the half-angle at the tip of the convergent θ_{cI} could then be greater than θ_{cO}.

In the limiting state in which the angle MOT is equal to ($\frac{\pi}{2} - \phi_p$), after projection on GT and OT we obtain:

on GT $\gamma \sin \theta_{cI} = R \cos \phi_p + F$

on OT $\gamma \cos \theta_{cI} = R \sin \phi_p$

Hence, eliminating R:

$$\gamma \cos \theta_{cI} \cot \phi_p = \gamma \sin \theta_{cI} - F$$

This equation defines θ_{cI} by taking the square and eliminating $\cos \theta_{cI}$. In fact, $\cos^2 \theta_{cI} = 1 - \sin^2 \theta_{cI}$. Let us recall that F can be calculated using Ergun's equation (see 1.7.2 T7).

If F becomes sufficiently large so as to reach the value of $F = \gamma$, the root θ_{cI} of the previous equation is equal to $\frac{\pi}{2}$. In other words, the air blown across a porous body fluidizes the product. This system is utilized in shipping tanks for pneumatic conveying in a dense phase.

On the other hand, the injection of air can reduce packing, that is, the consolidation of the product and, consequently, the unconfined stress limit f_c. This prevents arching.

The consumption of compressed air is of the order of 10 $Nm^3.h^{-1}$ at a relative pressure of 0.1–0.3 bar.

3.9.4. Forced emptying

Forced emptying is different to the injection of air via inclined walls since the air is now blown from top to bottom and its thrust F adds to the weight of the solid.

The apparent specific gravity then becomes:

$$\gamma' = \gamma + F$$

The increase in γ results in a reduction in the critical size of the outlet for withdrawal, and this disposition is sometimes utilized in the design of shipping tanks in a dense phase. In fact, the diameter of the outlet cannot be greater than that of the transport line. On the other hand, the product settles and is consolidated, especially if it is highly compressible (powder), and f_c increases. The forced withdrawal is therefore to be avoided.

3.9.5. Destabilization of the arch or dome by air knocker

The rapid opening of a capacity full of compressed air (at 6 bar rel. for example) linked to the hopper causes the arrival of air at the speed of sound, but this speed reduces very rapidly at the contact with the solid. The gas pressure disrupts the blocked areas in less than a second and causes the flow to resume.

Some companies propose an "air gun" directly attached to the wall of the hopper.

3.9.6. Destabilization of the arch by airbags

Airbags and elastomers (usually butyl rubber) are placed on the internal side of the wall of the convergent. Inflation and deflation alternations of the airbags are sufficient for the disruption of dome.

3.9.7. Other systems without vibration

These systems are necessary when the product has a rigid (or elastic) domain extended like, for example:

– fibers, spangles, scales, shavings.

– oily or moist products, possessing cohesive properties. These products sometimes have the additional defect of adhering to the wall.

The possible arrangements are:

1) Internal agitation. In the hopper, we have screws and/or ribbons vertical, horizontal or inclined (along the wall) ribbons, which cause internal circulatory motion and prevent the formation of compact blocks.

2) Sweep of the bottom of the silo by horizontal chains equipped with scrapping teeth, or with radial scrapers mounted on a vertical axis.

3) Installation of a horizontal rotating table under the withdrawal opening. A ploughshare takes the product off the table. We can adjust the rotatory motion of the table as well as its vertical distance from the opening.

3.10. Caking

3.10.1. *Caking (massification)*

We will see that the humidity and temperature of the interstitial gas have a decisive influence on caking.

For food products in general as well as powders of plant products in particular, this action is complex as it is most likely followed by unfamiliar chemical reactions.

On the other hand, for crystals and especially those which are at least partially soluble in water, the basic parameter is the chemical "activity" of water in the saturated crystal solution. This activity is also equal to the relative humidity ε of the gas in equilibrium with the solution:

$$\varepsilon = P_v / \pi(t) = a_e$$

P_v: partial pressure of the water vapor in the gas (Pa)

$\pi(t)$: pure water vapor pressure at the temperature t (Pa)

Let us assume that high humidity is followed by a severe dryness:

1) The adsorbed vapor on the crystals dissolve them superficially and these crystals are coated with saturated liquor. The product then shows hygroscopic behavior.

2) The change from humid to dry causes the vaporization of water and microcrystals precipitate and "cement" the particles among each other.

Note in this regard that a "deliquescent" product (which "liquefies" by the adsorption of water) possesses a very low value of ε (0.15 for LiCl and 0.32 for CaCl$_2$). On the contrary, an "efflorescent" product is covered with anhydrous crystals and corresponds to ε close to 1 (ε = 0.93 for Na$_2$SO$_4$ 10H$_2$O). We will find several values of ε in Mullin' book (p. 294). Whitening chocolate is efflorescent.

The cooling of crystals causes condensation since the partial pressure P$_v$ of water in the gas is such that:

$$P_v > a_e \pi(t)$$

Warming then gives rise to vaporization and massification:

$$P_v < a_e \pi(t)$$

But the effect of temperature can be more direct.

Thus, if ice is present on the surface of the particles, it can melt imperceptibly under the effect of contact stresses (compression). This fusion can thus occur at a temperature of $-1°C$ and be followed by re-solidification if the temperature drops. Hence caking (which would not have occurred if the temperature had stayed, for example, around $-10°C$).

The pitch weakens at 32°C and therefore has to be stored at a temperature of the order of 25–27°C maximum if we want to prevent the mutual "welding" of the particles, and therefore the caking of the product.

The creep (slow deformation) of the adipic acid due to stresses occurs from 65°C and, if we do not cool this product before storing it, there will again be mutual "welding" of the particles.

The higher the value of I$_c$ the easier it is for the crystals to be rigidly connected. Let us note the following.

Other than their chemical nature, the shape and size of the particles also come into picture and through this the coordination number I$_c$, which is the number of contact points existing between each particle and its neighboring particles.

Shape of particles	I_c
Spheroidal of uniform size (unimodal size)	4–6
Spheroidal of two sizes (granules and fine) (bimodal size)	>20
Fibers	>10
Low sphericity and spread-graded size	>20

Table 3.2. *Coordination number of D.S*

3.10.2. *Prevention of caking: coaters*

The above shows that variations in temperature and humidity are to be avoided during storage.

If the massification is partial enough not to prohibit flowing outside the hopper, a little coarse crusher will reduce the clods, which can allow for pneumatic conveying.

Examples of coaters include:

– for the treatment of some fertilizers: kieselguhr, micronized limestone, talc;

– for the treatment of detergents: micronized limestone treated with fatty amines or fatty alcohol sulfates.

– for salt (NaCl): magnesium carbonate or calcium and aluminum double silicate.

– cattle feed: starch, micronized limestone.

Other coaters include kaolin, silica, and magnesia.

What is the action mechanism of the coaters and additives of flowability?

These additive represent less than 1% of the mass of the D.S. They can act in the following manner:

– adsorption of gas or vapor which, consequently, does not attach to the D.S. The size of this adsorbent is of the order of one hundred nanometers.

– effect of "lubricant" of additive particles between the particles of the D.S.

– possession of residual electric charges, which hinder the agglomeration of the D.S.

4

Mechanics of Divided Solids

4.1. Static limit of divided solids: method of characteristics

4.1.1. *Equation of the field of stresses*

Let us consider a coordinate system where Oy is directed toward the bottom and Ox towards the left. The LRS is the static rupture line to which Mohr's circle of center W is tangent.

From Mohr's circle, we have:

$$WS = p\sin\phi = WP$$

$$\sigma_{xx} = p + WP\cos2\beta - TO \qquad [4.1]$$

$$\sigma_{xx} = p(1 + \sin\phi\cos2\beta) - c \times \cotan\phi \qquad [4.2]$$

$$\sigma_{yy} = p(1 - \sin\phi\cos2\beta) - c \times \cotan\phi \qquad [4.3]$$

$$\tau_{yx} = -\tau_{xy} = p\sin\phi\sin2\beta \qquad [4.4]$$

Equations [4.1] and [4.2] are generally written as below, for equilibrium:

$$\frac{\partial\sigma_{xx}}{\partial x} + \frac{\partial\tau}{\partial y} = f_x = 0 \qquad [4.5]$$

$$\frac{\partial \sigma_{yy}}{\partial y} + \frac{\partial \tau_{yx}}{\partial x} = f_y = \gamma \qquad [4.6]$$

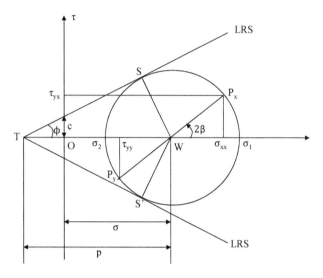

Figure 4.1. *Mohr's circle and LRS*

The x-axis is horizontal and the y-axis is vertical.

γ is the specific gravity of the D.S. (divided solid).

Now let us find the partial derivatives:

$$\frac{\partial \sigma_{xx}}{\partial x} = \frac{\partial p}{\partial x}(1 + \sin\phi\cos2\beta) - 2p\sin\phi\sin2\beta\frac{\partial \beta}{\partial x} \qquad [4.7]$$

$$\frac{\partial \tau}{\partial y} = \frac{\partial p}{\partial y}\sin\phi\sin2\beta + 2p\sin\phi\cos2\beta\frac{\partial \beta}{\partial y} \qquad [4.8]$$

$$\frac{\partial \sigma_{yy}}{\partial y} = \frac{\partial p}{\partial y}(1 - \sin\phi\cos2\beta) + 2p\sin\phi\sin2\beta\frac{\partial \beta}{\partial y} \qquad [4.9]$$

$$\frac{\partial \tau}{\partial x} = \frac{\partial p}{\partial x}\sin\phi\sin2\beta + 2p\sin\phi\cos2\beta\frac{\partial \beta}{\partial x} \qquad [4.10]$$

Let us substitute these derivation in the equilibrium equations [4.5] and [4.6]:

$$\left(1+\sin\phi\cos2\beta\right)\frac{\partial p}{\partial x} - 2p\sin\phi\sin2\beta\frac{\partial\beta}{\partial x} + \sin\phi\sin2\beta\frac{\partial p}{\partial y}$$

$$+ 2p\sin\phi\cos2\beta\frac{\partial\beta}{\partial y} = 0$$

[4.11]

$$\sin\phi\sin2\beta\frac{\partial p}{\partial x} + 2p\sin\phi\cos2\beta + \left(1-\sin\phi\cos2\beta\right)\frac{\partial p}{\partial y}$$

$$+ 2p\sin\phi\sin2\beta\frac{\partial\beta}{\partial y} = \gamma$$

[4.12]

Let us eliminate $\partial p/\partial y$ by multiplying equation [4.11] by $(1-\sin\phi\,\cos2\beta)$ and multiplying equation [4.12] by $\sin\phi\,\sin2\beta$. Subtracting corresponding terms:

$$\cos^2\phi\frac{\partial p}{\partial x} = -2p\sin\phi\left[\frac{\partial\beta}{\partial x}\sin2\beta + \frac{\partial\beta}{\partial y}\left(\sin\phi - \cos2\beta\right)\right] - \gamma\sin\phi\sin2\beta \qquad [4.13]$$

Similarly, after elimination of $\partial p/\partial x$:

$$\cos^2\phi\frac{\partial p}{\partial y} = -2p\sin\phi\left[\frac{\partial\beta}{\partial x}\left(\sin\phi + \cos2\beta\right) + \frac{\partial\beta}{\partial y}\sin2\beta\right] + \gamma\left(1+\sin\phi\cos2\beta\right) \qquad [4.14]$$

4.1.2. Total derivatives of β and p

Let us examine equations [4.2], [4.3] and [4.4]. If we have ϕ and c, we see that:

– the value of the angle β determines the direction of the stress t since Mohr's circle is tangent to the LRS;

– the value of p gives the magnitude of the stress $\vec{t} = \vec{\sigma} + \vec{\tau}$ acting on a surface whose normal makes an angle β with the major principal direction $o\sigma_1$ (see section 4.2.6).

Let us introduce the angle ζ between dx and dβ:

$$\zeta = (dx, \ d\beta)$$

The total derivative of β is:

$$d\beta = \left(\frac{\partial \beta}{\partial x}\cos\zeta + \frac{\partial \beta}{\partial y}\sin\zeta\right)ds \qquad [4.15]$$

But, for every vector V of the plane, the change in its magnitude is perpendicular to the change in its direction. This change in the magnitude of the vector is proportional to dp.

The angle between dx and dp is:

$$(dx, \ dp) = (dx, \ d\beta) + (d\beta, \ dp) = \zeta + \frac{\pi}{2}$$

Consequently:

$$dp = \left[\frac{\partial p}{\partial x}\cos\left(\zeta + \frac{\pi}{2}\right) + \frac{\partial p}{\partial y}\sin\left(\zeta + \frac{\pi}{2}\right)\right]ds$$

or:

$$dp = \left(-\frac{\partial p}{\partial x}\sin\zeta + \frac{\partial p}{\partial y}\cos\zeta\right)ds \qquad [4.16]$$

Nedderman [NED 92] did not give these explanations and his equation [7.6.5] is false. Nonetheless, without mentioning it, he uses our equation [4.16] and has thus obtained accurate results.

NOTE.–

In Figure 4.4, we see that the angle $(0\sigma, \ 0\tau)$ is equal to $-\dfrac{\pi}{2}$, which is why we have not written:

$$(dp, \ d\beta) = \frac{\pi}{2} \quad \text{and the exact angle is}: \ (d\beta, \ dp) = \frac{\pi}{2}$$

4.1.3. *Characteristic curves and their equations*

Let us substitute the values of $\partial p/\partial x$ and $\partial p/\partial y$ given by the equations [4.13] and [4.14] in equation [4.16]:

$$\frac{dp}{ds}\cos^2\phi = 2p\sin\phi\left[\sin(2\beta-\zeta)-\sin\phi\sin\zeta\right]\frac{\partial\beta}{\partial x}$$
$$+2p\sin\phi\left[\sin\phi\cos\zeta-\cos(2\beta-\zeta)\right]\frac{\partial\beta}{\partial y} \qquad [4.17]$$
$$+\gamma\left[\sin\zeta-\sin\phi\sin(2\beta-\zeta)\right]$$

We can eliminate $\dfrac{\partial\beta}{\partial x}$ from [4.17] by using [4.15]. Equation [4.17] becomes:

$$\frac{dp}{ds}\cos^2\phi = 2p\sin\phi\left[\sin(2\beta-\zeta)-\sin\phi\sin\zeta\right]\frac{1}{\cos\zeta}\frac{d\beta}{ds}$$
$$+2p\sin\phi\left\{\sin\phi\cos\zeta-\cos(2\beta-\zeta)-\left[\sin(2\beta-\zeta)-\sin\phi\sin\zeta\right]\tan\zeta\right\}\frac{\partial\beta}{\partial y} \qquad [4.18]$$
$$+\gamma\left[\sin\gamma-\sin\phi\sin(2\beta-\zeta)\right)$$

If we take out the coefficient of $\partial\beta/\partial y$, we would be left with derivatives with respect to s in equation [4.18]. The result can be written as:

$$\cos(2\beta-2\zeta)=\sin\phi=\cos2\varepsilon \qquad [4.19]$$

The angle ε is defined by:

$$\varepsilon=\frac{\pi}{2}-\phi$$

From [4.19]:

$$\beta-\zeta=\pm\varepsilon=\pm\left(\frac{\pi}{4}-\frac{\phi}{2}\right)$$

$$\tan\zeta=\frac{dy}{dx}=\tan(\beta-\varepsilon) \quad \text{or} \quad \tan\zeta=\frac{dy}{dx}=\tan(\beta+\varepsilon) \qquad [4.20]$$

Thus we obtain two directions ζ perpendicular to the direction $\zeta + \dfrac{\pi}{2}$, which is the direction dp/ds of the rupture stress. In other words, the inclinations ζ are those of the sliding planes.

The equation [4.18] is now written as:

$$\frac{dp}{ds}\cos^2\phi = 2p\sin\phi\left[\sin\left(2\beta-\zeta\right)-\sin\phi\sin\zeta\right]\frac{1}{\cos\zeta}\frac{d\beta}{ds}$$
$$+\gamma\left[\sin\zeta-\sin\phi\sin\left(2\beta-\zeta\right)\right] \tag{4.21}$$

Let us now use the fact that:

$$\frac{dx}{ds}=\cos\zeta=\cos\left(\beta\pm\varepsilon\right) \quad\text{and}\quad \frac{dy}{ds}=\sin\left(\beta\pm\varepsilon\right) \tag{4.22}$$

The equations of the two characteristics become:

$$\frac{dp}{ds}-2p\tan\phi\frac{d\beta}{ds}=\gamma\left[\frac{dy}{ds}-\tan\phi\frac{dx}{ds}\right] \tag{4.23a}$$

$$\frac{dp}{ds}+2p\tan\phi\frac{d\beta}{ds}=\gamma\left[\frac{dy}{ds}+\tan\phi\frac{dx}{ds}\right] \tag{4.23b}$$

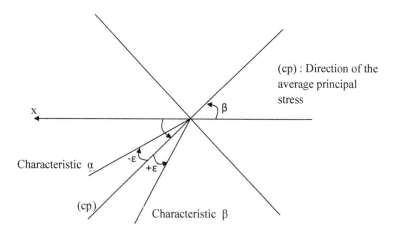

Figure 4.2. *Provision for the characteristics*

By multiplying the four terms of each equation [4.23] by ds, we obtain:

$$dp \pm 2ptan\phi d\beta = \gamma(dy \pm tan\phi dx)$$ [4.24]

In equations [4.22] and [4.24], we can replace dx, dy, dβ, and dp by differences by involving the values of x_1, y_1, and β_1 and p_1 relative to the point M_1 and the values of x_2, y_2, β_2, and p_2 relative to the point M_2.

$$y - y_i = (x - x_i)tg(\beta \pm \varepsilon)$$ [4.25]

$$p - p_i \pm 2tg\phi p_i[\beta - \beta_i] = \gamma(x - x_i)[tg(\beta \pm \varepsilon) \pm tg\phi]$$ [4.26]

The index i = 1 carries the + sign and the index i = 2 carries the – sign.

Thus, knowing the points M_1 and M_2, it is easy to determine the characteristics of the point M(x, y, β, p), which is the point of intersection of the characteristics passing through the points M_1 and M_2 (see Figure 4.3).

Let us take the example of a downward vertical direction for the major principal stress.

The two equations [4.25] and [4.26] enable us to determine the values of x, y, p, and β at the point M.

We thus determine, step by step, the characteristics of the stresses.

NOTE.–

It is also possible to determine a network of characteristics for displacements (see Nedderman [NED 92]).

Thus, we can describe:

– Mohr's circle for displacements or rather their velocity;

– a plastic potential;

– the application of the principle of coaxiality.

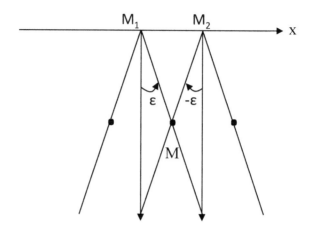

Figure 4.3. *Characteristic lines M_1M and M_2M*

4.2. The dynamic of D.S. according to Bagnold [BAG 54]

4.2.1. A divided solid subjected to a pure shear

The D.S. is assumed to be without cohesion. Bagnold [BAG 54] proves that, for a shear stress, the pressure between two parallel layers (of particles) is (his formula 6):

$$P = K_1 d_p^2 \left(\frac{dU}{dy} \right)^2$$

$U(y)$: local velocity of a layer (m.s^{-1})

y: perpendicular coordinate to the velocity U (m)

d_p: diameter of supposedly spherical particles (m)

The shear stress is:

$$T = K_2 d_p^2 \left(\frac{dU}{dy} \right)^2$$

P and T: stresses (Pa)

Bagnold [BAG 54] deduced, by means of his equation 9, an expression of the local velocity:

$$U = \frac{K_3}{d_p} y^{3/2}$$

The author indicates that the porosity is constant across the thickness of the moving bed.

But, the numerical simulation by Siebert *et al.* [SIE 01] showed that:

– the porosity is constant across the thickness of the bed;

– the velocity increases almost linearly with the distance from the bottom of the bed.

This confirms the arguments made by Bagnold [BAG 54].

4.3. Dynamics of divided solids: method of discrete elements (of distinct particles)

4.3.1. *Presentation*

Cundall and Strack [CUN 79] give details about the method of distinct particles. However, they limit it to two dimensions. We will generalize it to three dimensions. The procedure of numerical integration by Cundall and Strack [CUN 79] derive from that of Verlet [VER 67, VER 68]. First, we will examine the interparticle contact with adherence in detail and in the absence or presence of a force located in the plane tangent to the two particles in contact (see section 4.3.2).

We will then examine the repulsive action existing between two particles that are not in contact (see section 4.3.4).

The relative movement of two particles will be described in three dimensions. The damping characterizing this movement has been described by Cundall and Strack [CUN 79]. We will specify this concept (see section 4.3.6).

The particles involved in the calculations are generally supposed to be spherical. We will determine the moment of inertia of a sphere in section 4.3.7. It generally suffices to vary their mutual coefficient of friction in order to account for nonspherical nature and nesting of the particles.

Potapov and Campbell [POT 98], on the one hand, study the general case of non-spherical particles. On the other hand, Reberties and Sando [REB 77] study the dynamics of the sphero-cylinders. Ting *et al.* [TIN 93] also study the effect of the shape of the particles.

A method, in itself, is described to select the particle couples close enough for their forces of interaction to be significant (see section 4.3.9).

The integration of the movement of particles is done by increments in time which must be:

– not too large to ensure the stability of the calculation;

– not too small in order to save calculation time.

Ting and Corkum [TIN 92] provide an example of the application of the method of discrete elements.

Bathurst and Rothenburg [BAT 88] as well as Chang *et al.* [CHA 89] propose the method for deducing the stress tensor from the microscopic structure of a divided solid.

4.3.2. Contact between two particles

1) Contact with adherence without tangential component of force.

Without adherence, Hertz's classic relationship can be written as:

$$a^3 = \frac{F_N R^*}{K} = \frac{3F_N R^*}{4E^*}$$

a: radius of the circular contact surface (m)

F_N: force pressing both spheres one against the other (N)

$$\frac{1}{R^*} = \frac{1}{R_1} + \frac{1}{R_2}$$

$$\frac{1}{K} = \frac{3}{4}\left(\frac{\left(1-v_1^2\right)}{E_1} + \frac{\left(1-v_2^2\right)}{E_2}\right) = \frac{3}{4E^*}$$

R_1 and R_2: radii of curvature of the contact surfaces (m)

E_1 and E_2: modulus of elasticity of particles in contact (Pa)

v_1 and v_2: Poisson's modulus for both particles: dimensionless

If there is adherence, the adhesion energy is defined by (see [4.22]):

$$\gamma = \gamma_1 + \gamma_2 - \gamma_{12}$$

γ_1 and γ_2: surface energies of both particles in the presence of their respective saturated vapors $(J.m^{-2})$

γ_{12}: interfacial energy of both particles in contact $(J.m^{-2})$

In the presence of adherence, the radius a_a of the contact circle (see Sankoor and Briggs [SAN 73]; Barquin *et al.* [BAR 77]; Kendall [KEN 86]) is written as:

$$a_a^3 = \frac{R^*}{K}\left\{F_N + 3\gamma\pi R^* + \left[6\gamma\pi R^* F_N + \left(3\gamma R^*\right)^2\right]^{1/2}\right\} = \frac{R^* F_{N1}}{K} \qquad [4.27]$$

According to Johnson *et al.* [JOH 71], the value F_S of F_N (negative) for the separation corresponds to $a_a = 0$

$$F_S = -\frac{3\pi\gamma R^*}{2}$$

2) Presence of a tangential force F_T.

Such a force, called tangential force, causes a relative displacement δ of both particles and without deformation and without changing the size of the contact circle. According to Savkoor and Briggs [SAV 77]:

$$\delta = \frac{F_T}{8a_{aT}G^*} \quad \text{with} \quad \frac{1}{G^*} = \frac{2-v_1}{G_1} + \frac{2-v_2}{G_2}$$

G_1 and G_2: shear moduli of particles 1 and 2 (Pa)

$$G_i = \frac{E_i}{2(1+v_i)}$$

Thornton [THO 91b] introduces the normal effective force F_{Ne} given by the equation:

$$F_{Ne} = F_{N1}\left[1 - \frac{(F_{N1} - F_N)}{3F_{N1}}\right]^{3/2}$$

F_{N1} is given by its equation [4.27].

From which he deduced the criterion for sliding

$$F_{TC} = \mu F_{Ne}\left[1 - \left(1 - \frac{\delta}{\delta_{max}}\right)^{3/2}\right]$$

or:

$$\left(1 - \frac{F_{TC}}{\mu F_{Ne}}\right)^{2/3} = \left(1 - \frac{\delta}{\delta_{max}}\right)$$

(see Mindlin, equation 201 [MIN 49] and Mindlin and Deresiewicz, equation 7 [MIN 53])

δ is the relative displacement of both particles (always tangent to the sliding plane) when the "peeling" is done and the sliding actually appears. The "peeling" phase exists when F_T varies from zero to F_{TC}.

$$\delta_{max} = \frac{3\mu F_{Ne}}{16G^* a_{aT}}$$

with

$$a_{aT}^3 = \frac{R^* F_{Ne}}{K} \qquad \left(\text{see Langston } et\ al.\ [\text{LAN 95}]\right)$$

Eliminating a_{aT} from these two equations (see Langston et $al.$'s Appendix [LAN 95])

$$\delta_{max} = \frac{3\mu F_{Ne}^{2/3}}{16G^*}\left(\frac{K}{R^*}\right)^{1/3} \quad \text{with} \quad \frac{1}{K} = \frac{3}{4E^*}$$

If we assume that both particles are made of the same material

$$\frac{1}{G^*} = \frac{4(1+v)(2-v)}{E}$$

EXAMPLE 4.1.–

We will examine the contact of two spherical particles superimposed by a diameter of 1 mm, where one of the two particles weighs on the other:

$d_p = 0.001$ m $\gamma = 0.02$ J.m^{-2} $v = 0.3$

$\rho_s = 2700$ kg.m^{-3} $E_i = 10^5$ Pa $\mu = 0.5$

$$F_N = \frac{2700 \times \pi \times 10^{-9}}{6} = 1.413 \times 10^{-6}\,N$$

$$R^* = \frac{1}{2}\frac{d_p}{2} = 0.25 \times 10^{-3}\text{ m}$$

$$F_{N1} = 1.413 \times 10^{-6} + 3 \times 0.02 \times \pi \times 0.25 \times 10^{-3}$$

$$+ \left[6 \times 0.02 \times \pi \times 0.25 \times 10^{-3} \times 1.413 \times 10^{-6} + \left(3 \times 0.02 \times \pi \times 0.25 \times 10^{-3}\right)^2\right]^{1/2}$$

$$F_{N1} = 48.54 \times 10^{-6} + \left[1.33 \times 10^{-10} + 22 \times 10^{-10}\right]^{1/2} = (48.54 + 48.30) \times 10^{-6}$$

$$F_{N1} = 9.684 \times 10^{-5}\,N$$

$$\frac{1}{K} = \frac{3}{4}\left(\frac{1-0.09}{10^5} + \frac{1-0.09}{10^5}\right) = 1.365 \times 10^{-5}\,Pa^{-2}$$

$$a_a = \left[0.25 \times 10^{-3} \times 9.684 \times 10^{-5} \times 1.365 \times 10^{-5} \right]^{1/3}$$

$$a_a = 0.69 \times 10^{-4} \text{ m}$$

$$F_{Ne} = 9.684 \times 10^{-5} \left[1 - \frac{9.684 \times 10^{-5} - 0.1413 \times 10^{-5}}{3 \times 9.684 \times 10^{-5}} \right]^{3/2}$$

$$F_{Ne} = 9.684 \times 10^{-5} \left[1 - 0.3284 \right]^{3/2} = 5.33 \times 10^{-5} \text{ N}$$

$$\frac{1}{G^*} = \frac{4(2 - 0.3) \times (1 + 0.3)}{10^5} = 8.84 \times 10^{-5}$$

$$\delta_{max} = \frac{3 \times 0.5 \times 5.33 \times 10^{-5} \times 8.84 \times 10^{-5}}{16 \times 0.15 \times 10^{-3}} = 7 \times 10^{-9} \text{ m}$$

$$\delta_{max} = 7 \text{ nm}$$

This displacement for the "peeling" is very low, which justifies the use of the simple expression using an empirical value μ_E of μ.

$$F_T = \mu_E F_{NE}$$

Here, F_{NE} just as F_T only results from the movement of particles as well as the movement of an external field (if it exists).

4.3.3. Interparticle friction according to Cundall and Strack [CUN 79]

These authors apply Coulomb's law when sliding occurs

$$F_T = C + \mu F_N$$

C: cohesion (Pa)

μ: empirical factor of mutual friction: dimensionless

F_N: normal force to the sliding plane (N)

F_T: force called "tangential" force (N)

The coefficient μ increases with:

– the surface roughness of the particles;

– the nonsphericity of the particles which have cavities, bumps, sharp tips, crests.

The effect of these properties has never been studied.

Recall that the nonsphericity is the ratio of the surface areas of the particle to the surface area of the particle of the sphere of equal volume.

Note that, when the frictions (and the cohesion) between particles cancel out and the size of the particles reduces, the D.S. draws closer to being a liquid.

4.3.4. Forces of interaction between two particles

In calculations regarding distinct particles, the force generally considered by researchers is a *repulsive* force.

1) The interaction continues from a potential ϕ:

$$F_{TC} = \frac{-\partial\phi}{\partial r} = \frac{n\varepsilon\left(\dfrac{\sigma}{r}\right)^n}{r} \quad \text{with} \quad n = 36$$

ε: energy parameter (J)

2) Elastic interaction:

$$F_{el} = k(\sigma - r)$$

σ: diameter of two equal superimposed particles (else $\sigma = R_1 + R_2$)

r: distance between centers of both particles

We notice that if $r < \sigma$, there is interpenetration of one particle in the other (overlap). But, in reality, the two surfaces in contact flatten and the contact surface is a circle of radius a called contact circle.

3) Hertz's interaction [HER 81]:

$$F_{Hz} = \frac{4}{3} E^* \sqrt{R^*} \left(\sigma - r\right)^{3/2} \quad \left(\text{Langston } et \ al. \ [LAN\ 95]\right)$$

$$\frac{1}{R^*} = \frac{1}{R_1} + \frac{1}{R_2} \qquad \frac{1}{E^*} = \frac{1 - v_1^2}{E_1} + \frac{1 - v_2^2}{E_2} \quad \left(\text{Thornton } [THO\ 91b]\right)$$

$$a^3 = \frac{3\pi}{4} R^* \left(k^*\right)^2 P_o \qquad k^* = \frac{1 - v^2}{\pi E^*} = k_1 + k_2$$

The approximation δ between both surfaces is given by Johnson $et \ al.$ [JOH 71]

$$\delta^3 = \frac{9\pi^2}{16} \frac{\left(k^*\right)^2 P_o^2}{R^*} \quad \left(\text{assuming } v_1 = v_2 = v\right)$$

We notice with Johnson [JOH 58] a tensile stress on the periphery of the contact circle, a compressive stress at the center of the contact circle.

This has been confirmed by Kendall [KEN 86].

4.3.5. Movement at the contact point between two particles i and j

1) Relative movement of both particles parallel to the axis of their centers.

For this purpose, let us introduce the distance (positive by definition) between the centers of both particles. Let D_{ij} be this distance. The direction cosines of the direction $\bar{n}$ from i toward j are:

$$\alpha_{nij} = \frac{x_j - x_i}{D_{ij}} \qquad \beta_{nij} = \frac{y_j - y_i}{D_{ij}} \qquad \gamma_{nij} = \frac{z_j - z_i}{D_{ij}}$$

Both particles being in contact:

$$D_{ij} = r_i + r_j$$

x_i, y_i, and z_i are the coordinates of the center point M_i and x_j, y_j, and z_j are the coordinates of the center M_j.

The components of the relative movement from center to center relative to the lines of center are:

$$\left(\dot{x}_j - \dot{x}_i\right)\alpha_{nij} \quad \left(\dot{y}_j - \dot{y}_i\right)\beta_{nij} \quad \left(\dot{z}_j - \dot{z}_i\right)\gamma_{nij} \quad \text{but} \quad \vec{n}.\left(\vec{M}_j - \vec{M}_i\right)$$

The point above the letters indicate a derivative with respect to time.

2) The relative movement of both particles parallel to both surfaces in contact, that is perpendicular to the line of centers.

$$\vec{v}_i - \vec{v}_j = \vec{\omega}_i \times \vec{n}_i r_i + \vec{\omega}_j \times \vec{n}_i r_j = \left(\vec{\omega}_i r_i + \vec{\omega}_j r_j\right) \times \vec{n}_i = \overrightarrow{\Omega R} \times n_i$$

The external (or vector) product of two vectors (represented by the symbol $\times$) is described in Appendix 3.

$\vec{\omega}_i$ and $\vec{\omega}_j$ define the pivoting speed counted positively (in the anticlockwise direction) at the foot of each vector.

r_i and r_j are the radii of the particles i and j assumed to be spherical.

The symbol $\times$ represents an external or vector product.

The unit vector $\vec{t}$ is:

$$\vec{t} = \frac{\overrightarrow{\Omega R} \times \vec{n}_i}{\left|\Omega R\right|}$$

The direction cosines of $\vec{t}$ will be known as α_{tij}, β_{tij}, and γ_{tij}. They will be easy to find knowing ω_{ix}, ω_{iy}, ω_{iz}, ω_{jx}, ω_{jy}, and finally ω_{jz}.

The tangential components (perpendicular to the line of centers) of the displacement of M_iM_j are therefore:

$$\alpha_{tij}\left(\dot{x}_i - \dot{x}_j\right) + \beta_{tij}\left(\dot{y}_i - \dot{y}_j\right) + \gamma_{tij}\left(\dot{z}_i - \dot{z}_j\right)$$

That is the scalar product (we also call it interior) represented by a dot on the line:

$$\vec{t}.\overrightarrow{M_iM_j} = \dot{s}$$

By adopting the notations of Cundall and Strack [CUN 79], we obtain:

$$\overrightarrow{\Delta n} = \dot{\vec{n}}\,\Delta t = \overrightarrow{M_iM_j}.\vec{n}\Delta t$$

$$\overrightarrow{\Delta s} = \dot{\vec{s}}\,\Delta t = \left\{\overrightarrow{M_iM_j}.\vec{t} + \overrightarrow{\Omega R} \times \vec{n}_i\right\}\Delta t$$

These are Cundall and Strack's equations 14 and 15 [CUN 79].

The dot on the line here represents the interior product (or scalar product) of two vectors.

In their publication, Tsuji *et al.* [TSU 92] give useful specifications on the impact of particles on a wall.

4.3.6. *Damping*

According to Cundall and Strack [CUN 79], the equations of motion are:

$$m_1\,\ddot{x}_1 = \Sigma\left[F_{1x} + D_{1x}\right] - C\dot{x}_{1x}$$

$$I_1\,\ddot{\theta}_1 = \Sigma M_1 - C^*\dot{\theta}_1$$

The coefficients C and C^* here define the damping respectively for the translation and rotation around an axis passing through the particle's center of mass.

I: moment of inertia of the particle

M: couple of force acting on the particle

m: mass of the particle

NOTE.–

The damping called as "contact damping" of Cundall and Strack [CUN 79] appeared to be useless.

NOTE.–

Silbert *et al.* [SIL 01] readopt Cundall and Strack's logic [CUN 79] by improving it and assuming it in three dimensions.

4.3.7. Moment of inertia of a sphere

This quantity comes into play when expressing the pivot energy of a particle with respect to an axis passing through its center of mass (the barycenter).

The moment of inertia is defined by:

$$I = \int_0^m r^2 dm$$

m: mass of the particle

r: distance from a point of the particle to the axis

A very important particular case is the moment of inertia of a sphere with respect to an axis passing through its center.

$$I = \frac{2}{5}mR^2 \quad \left(\text{Ferry } et \text{ } al. \text{ } [\text{FER } 81, \text{p.}138]\right)$$

m: mass of the sphere (kg)

R: radius of the sphere (m)

4.3.8. Value of the increment in time

1) According to Rong *et al.* [RON 95].

Table 2 of [RON 95] gives the elastic stiffness K_n in $kN.m^{-1}$ and not in $kN.m^{-2}$

Equation 18 gives (p. 251):

$$\Delta t_c = \frac{\pi}{\sqrt{\dfrac{K_n}{m}\left(1 - \dfrac{Ln^2 e}{\pi^2 + Ln^2 e}\right)}}$$

where e is the coefficient of restitution

EXAMPLE 4.2.–

e = 0.95 ρ=2700 kg.m^{-3}

$d_p = 10^{-3}$ m $K_n = 1.5 \times 10^5$ N.m^{-1} (ou $\dfrac{kg}{s^2}$)

$$m_p = \frac{\pi.10^{-9}}{6} \times 2700 = 1.41 \times 10^{-6} \text{ kg}$$

$$\Delta t_c = \frac{3.1416}{\left[\dfrac{1.5 \times 10^5}{1.41 \times 10^{-6}}\left(1 - \dfrac{2.63}{9.86 + 2.63}\right)\right]^{1/2}} = 1.08 \times 10^{-5} \text{ s}$$

2) According to Langston $et\ al.$ [SEL 95]

a) Equation 16 on page 981 gives:

$$\Delta t_c = \left(\frac{\pi R}{\alpha}\right)\left(\frac{\rho}{G}\right)^{1/2}$$

But, for glass and iron, the ratio E/ρ = (2.6 ± 0.1) $\times 10^7$ m^2.s^{-2}:

$$\frac{G}{\rho} = \frac{E}{2(1+\upsilon)\rho} = \frac{2.6 \times 10^7}{2(1+0.3)} = 10^7 \text{ m}^2.\text{s}^{-2}$$

For example:

$R = 0.5.10^{-3}$ m $\dfrac{G}{\rho} = 5.37 \times 10^7$ s

$\alpha = 0.92$ rad

$$\Delta t_c = \frac{\pi \times 0.5 \times 10^{-3}}{0.92} \times 10^{-3,5} = 5.37 \times 10^{-7} \, s$$

b) Equation 15 on page 972 gives:

$$\Delta t_c = 0.1 \left(\frac{m}{k} \right)^{1/2}$$

For example:

$$m = 1.41 \times 10^{-6} \, kg \qquad k_n = 1.5 \times 10^7 \, N.m^{-1}$$

$$\Delta t_c = 0.1 \left(\frac{1.41 \times 10^{-6}}{1.5 \times 10^7} \right)^{1/2} = 3 \times 10^{-7} \, s$$

c) In practice, according to Thompson and Grest [THO 91a], we choose from their equations 15 and 16:

$$\left. \begin{matrix} \text{equ } 15 \\ \text{equ } 16 \end{matrix} \right\} \quad \Delta t = \Delta t_c \quad \left\{ \begin{matrix} = 50 \times 3 \, 10^{-7} = 1.5 \, 10^{-5} \, s \\ = 50 \times 5.37 \, 10^{-7} = 2.7 \, 10^{-5} \, s \end{matrix} \right.$$

d) We will find useful specifications concerning the increment in time in Tsuji *et al.*'s [TSU 92] publication.

4.3.9. Selection of influential particles on a particle

Let us assign to each particle a matriculation number i. Henceforth, each particle possesses the following properties:

– x_i, y_i, z_i: the three components of the position of its center;

– v_{xi}, v_{yi}, v_{zi}: the three components of the velocity of its center;

– $\omega_{xi}, \omega_{yi}, \omega_{zi}$: the three components of its pivot velocity.

Preselection of the particles of influence on the particle i.

1) We select the particles j, which are such that:

$$x_i - \Delta l < x_j < x_i + \Delta l$$

2) Among the particles j, we select the particles k such that:

$$y_i - \Delta l < y_k < y_i + \Delta l$$

3) Among the particles k, we select the particles ℓ such that:

$$z_i - \Delta l < z_l < z_i + \Delta l$$

Thus, the particles ℓ retained are contained in a cube of side 2 $\Delta \ell$ of which at the center the particle I is situated.

We can then have the radius of influence r_i such that among the particles of type ℓ, the particles of type m are such that:

$$\left(x_i - x_m\right)^2 + \left(y_i - y_m\right)^2 + \left(z_i - z_m\right)^2 < r_i^2 \quad \text{with} \quad r_i < \Delta l$$

Finally, the particles m retained as being able to influence the particle i are the particles contained in the sphere of radius r_i and whose center coincides with that of the particle i. These particles are the particles of influence on the particle i.

We have thus considerably reduced the number of particles capable of affecting the particle i. We now have to evaluate the interactions between the particle i and its particles of influence.

4.3.10. *More economical study of significant couples*

The system is partitioned into elementary cubic regions k each containing particles.

Each particle of a region k is related to the particles:

– of the region k;

– of contiguous regions to the region k.

Thus, we are not obliged to find the particles involving all the particles of the system.

Rong *et al.* [RON 95], in their Figures 4 and 5, describe the mode of operation.

4.3.11. *Coefficient of restitution*

The equations of motion of the point of contact are of the form:

$$F = m\ddot{x} + c\dot{x} + kx$$

The coefficient c can be replaced by the product mD.

The dimensional equation of the coefficients are as follows:

$$[F] = N = \frac{kg \times m}{s^2} \qquad\qquad [k] = \frac{N}{m} = \frac{kg}{s^2}$$

$$[c] = \frac{N \times s}{m} = \frac{kg}{s} \qquad\qquad [D] = \frac{kg}{s \times kg} = \frac{1}{s}$$

This second order linear equation has the following roots for its characteristic equation:

$$r = -\frac{c}{2m} \pm \sqrt{\left(\frac{c}{2m}\right)^2 - \frac{k}{m}} \quad < 0$$

The solution is therefore:

$$x = x_{o1}e^{r_1 t} + x_{o2}e^{r_2 t}$$

By keeping one of the roots r_i:

$$x = x_{oi}e^{r_i t} \qquad r_i < 0 \qquad \dot{x} = \dot{x}_{oi}e^{r_i t}$$

For the motion parallel to the line of centers, the existence of a to and fro movement justifies a coefficient 2:

$$t = \frac{2}{r_i}Ln\left(\frac{\dot{x}}{\dot{x}_{oi}}\right) = -\frac{2Lne}{r_i}$$

e is the coefficient of restitution.

In practice, the coefficient of restitution is an approximate value which helps in the calculation of the increment in time. We will find interest in reading the publication of Tsuji *et al.* [TSU 92] who establishes the relationship between the coefficient of restitution and the increment in time.

4.3.12. *Calculation of k and c*

According to Thompson and Grest [THO 91a], parallel to the line of centers:

$$k_n = 2 \times 10^5 \times \frac{mg}{d_p}$$

$$c_n = 80.5 \left(\frac{g}{d_p} \right)^{1/2}$$

For example:

$$m = 1.41 \times 10^{-6} \text{ kg} \qquad d_p = 10^{-3} \text{ m} \qquad g = 9.81 \text{ m.s}^{-2}$$

$$k_n = \frac{2 \times 10^5 \times 1.41 \times 10^{-6} \times 9.81}{10^{-3}} = 2760 \text{ kg.s}^{-2}$$

$$c_n = 80.5 \times \left(\frac{9.81}{10^{-3}} \right)^{1/2} = 7974 \text{ kg.s}^{-1}$$

According to Thompson and Grest [THO 91a]:

$$k_s = \frac{k_n}{2}$$

According to Cundall and Strack [CUN 79]:

$$c_s = c_n$$

NOTE.–

Tsuji *et al.* [TSU 92] describe the method of repeating these calculations from Hertz's assumption and not from the elastic hypothesis.

4.3.13. *Flow on an inclined plane by the method of discrete elements*

Siebert *et al.* [SIE 01] give a *detailed* example of the usage of the method of discrete elements (M.D.E.).

They make use of an assembly whose thickness is of one hundred particles, which is close to the physical reality and they give the plane an inclination greater than 20 deg.

1) Compactness profile β.

The *compactness* is the ratio of the solid volume to the total volume (solid + vacuum).

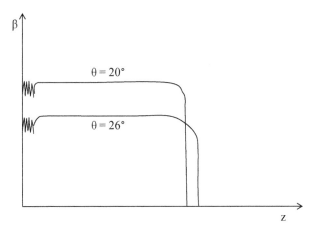

Figure 4.4. *Variation of the compactness profile with the distance z from the bottom*

If the slope increases, an expanded region increases close to the surface. Apart from this region, the compactness β *remains constant* along the flowing thickness. The compactness reduces linearly with the angle of slope.

2) We have two types of flow depending on the angle of inclination θ:

stable flow $\theta_r < \theta < \theta_{in}$

unstable flow $\theta_{in} < \theta$

The resting slope θ_r depends on the thickness H of the particle bed as well as the slope θ_{in} (see the authors' Figure 2).

3) Particular temperature profile.

The particular "temperature" is proportional to the average of the squares of the velocity fluctuations

$$T_p = \rho \langle \delta v^2 \rangle$$

T_p is maximum toward the base meanwhile the compactness β stays constant along the height of the flowing bed.

4) Flow rate.

This velocity increases almost proportionately with the distance z from the bottom and increases with the thickness of the flowing bed.

The velocity increases with the interparticle coefficient of friction μ. If μ reduces to 0.1, the flow becomes unstable.

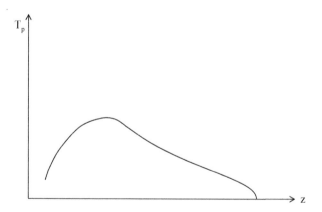

Figure 4.5. *Particulate temperature T_p depending on the distance z from the bottom*

5) Coordination number N_c.

When the slope increases from the angle of repose θ_r, N_c decreases from 4 to 0.5, that is from a tight compactness to a loose compactness.

NOTE.–

Kruyt and Verël [KRU 92] came up with a rather simple theory which gave to the velocity a slightly sinusoidal profile.

4.4. Surface dynamics of a D.S.

4.4.1. *Bouchaud et al.'s equation [BOU 94] simplified by de Gennes [DE 95, DE 97]*

We will involve the following quantities:

R: number of rolling particles per horizontal unit area

F: number of fixed particles (stationary) per horizontal unit area

ω: volume occupied by each fixed particle taking the immediate surrounding vacuum into consideration. This volume is such that:

$$\frac{\pi d^3}{6} = (1 - \varepsilon)\omega \quad \text{hence} \quad \omega = \frac{\pi d_p^3}{6(1 - \varepsilon)}$$

d_p: diameter of a particle (supposedly spherical) (m)

ε: fraction of void of fixed volume

h: height of the fixed portion. This height is such that:

$$\frac{hA}{\omega} = FA \quad \text{hence} \quad h = F\omega = \frac{\pi d_p^3 F}{6(1 - \varepsilon)}$$

A: bottom surface area of the system (m^2)

v: velocity (counted horizontally) of displacement of rolling particles. This velocity is counted positively downstream.

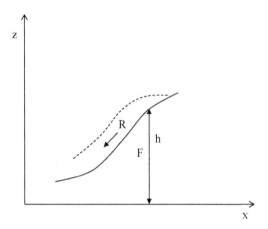

Figure 4.6. *Fixed and rolling particles*

Along the horizontal length Δx and during the time Δt, the change in the number of fixed (stationary) particles N is:

$$\Delta N = dF\Delta x = \gamma \Delta t \left(\theta - \theta_r\right) R\Delta x$$

θ_r: angle of repose (radian)

γ: exchange frequency between F and R ($s^{-1}.rad^{-1}$)

Let us divide by $\Delta x \Delta t$:

$$\frac{\partial R}{\partial t} = \gamma\left(\theta - \theta_r\right) R - v\frac{\partial R}{\partial x}$$

The first term shows the change of state from the state F to the state R. We can call this term "accretion term".

The second term is a convection term.

4.4.2. Application of de Gennes' equation [DE 97]

1) Boutreux and de Gennes [BOU 97] studied and solved the problem of the spreading of a step of sand. In order to achieve this, they came up with

the relationship relating the initial width of the walking sand and its final width. They also came up with the total duration of the spreading process of the walk.

2) Boutreux and de Gennes [BOU 98] showed that, if we pour a D.S. on a ridge of sand with slope less than the angle of repose θ_r, a sharper ridge of slope close to θ_r is formed on a horizontal half-width proportional to the square root of time.

NOTE.–

A snowy carpet subject to variations in gel and thaw becomes resistant (consolidated) because the flakes have become mutually interlocked.

If fresh snow falls on a consolidated carpet, the newly arrived flakes are not linked one on the other and an avalanche is therefore possible.

CONCLUSION.–

The knowledge acquired up to now enables us to exploit several results practically. However, it seems like a more systematic use of the effect of van der Waals forces (Hamaker's equations [HAM 34]) would improve the numerous results already obtained. It would probably help to make use of a force derived from the attractive part of Lennard–Jones's potential at least for fine powders:

$$V_{att} = -4\varepsilon \left(\frac{\sigma}{r} \right)^6 \quad \text{hence} \quad F = \frac{24\varepsilon}{r} \left(\frac{\sigma}{r} \right)^6$$

The comparison of these two ways of expressing attraction would enable us to determine the energy coefficient ε.

4.4.3. Side of a layer of D.S. analytical study

Zang and Foda [ZAN 97] suggest a mathematical solution to the problem of avalanches (especially that of snow).

4.5. Experimental studies

Based on his industrial experience, Johanson [JOH 79] explains some behaviors of gas–solid mixtures.

Carson and Marinelli [CAR 94] explain how to take the properties of a D.S. into consideration to obtain a regular flow.

During a horizontal pneumatic conveying process, Tsuji *et al.* [TSU 92] make use of the modified M.D.E. to calculate the concentration in the pipeline.

Spillmann *et al.* [SPI 05] study the couple of a paddle stirrer immersed in a divided solid.

Campbell *et al.* [CAM 85] studied the flow regimes of a noncohesive D.S. in an inclined chute. They noticed the existence of a regime said to be "laminar" with a thick bed and a "turbulent" regime of smaller width and greater velocity.

5

Densification of Powders: Tablets and Granules

5.1. Useful properties of powders for pressing

5.1.1. History (pharmaceutical powders)

Over the years, there have been several ways to easily transport predetermined amounts of powder:

– cachets were first obtained by bringing together the edges of two cupules of unleavened bread. One of the cupules was initially filled with an amount of powder. These cachets are no longer used;

– capsules are obtained by interlocking two cylindrical and oblong capsules and are made of hard gelatin. One of the two contains the powder. In order to fill it, we use a dosing tube where the powder is consolidated by a piston and ejected into one capsule. A sufficiently cohesive powder is required;

– tablets are prepared by pressing in a matrix by a piston (punch). The pressure can vary from 10 to 500 MPa, and more often between 50 and 300 MPa. Tablets generally have between 5 and 10 mm of diameter;

– suppositories are molded but are only used in France. They contain a large percentage of lubricant, likely to soften to the human body temperature.

Capsules, tablets and suppositories have numerous advantages:

– dosing accuracy;

– easy to use;

– storage stability;

– reduced cost for storage and transportation.

5.1.2. Industrial powders (conditioning)

Cement, gypsum and fertilizers are presented in hard paper bags. Salt is either used in bulk (roads, electrolysis), or conditioned in a box for domestic use (just as powdered sugar).

Flours and powdered plastics (before use) are also conditioned in paper bags.

5.1.3. Flowability of a powder

For the filling of the matrix of a tablet, it is important for the flowability to be good.

Measurements of the flowability with a shear cell (Jenike's cell or annular cell) will not be described here. This flowbility is especially used for the storage assessment of hoppers and silos.

Carr [CAR 65a] suggests a much simpler assessment of the flowability. Apres having filled a measuring cylinder with powder, we gently and regularly tap it with, for example, a wooden tool. The powder slowly settles up to a certain limit.

The initial apparent mass density of the powder is:

$$\rho_{a0} = M/V_0$$

M: mass of powder (kg)

V_0: initial volume of the powder (m^3)

The final limiting mass density:

$$\rho_{af} = M/V_f$$

V_f: final volume of the powder (m^3)

We then determine the "compressibility" as a percentage by:

$$C = \frac{\rho_{af} - \rho_{a0}}{\rho_{af}} \times 100$$

According to Carr [CAR 65b], the flowability is given by Table 5.1.

C	Flowability
5–15	Excellent
12–16	Good
18–21	Fair
23–25	Poor
33–38	Bad
> 40	Very bad

Table 5.1. *Compressibility and flowability*

Nyqvist [NYQ 84] compiles the measurements of flowability:

– the unconfined stress f_c;

– the flow rate across a calibrated orifice;

– the angle of repose.

The results obtained by Nyqvist [NYQ 84] show that the stress f_c is the most sensible property but the method of calibrated orifice is the fastest. The frequency of the necessary settings for the machine is directed related to the measurement of flowability by the calibrated orifice.

The use of powders is directly related to their flowability, which will be particularly bad if the particles have the tendency to get interlocked one with the other. For this, it suffices for example that, among the three principal directions, one of them be very different from the others (platelets, rods).

The roughness of the surface of the particles can also obstruct the flow but, at the same time, hinder a segregation that would be due to large differences in mass density. A thorough study of the mixture consisting of the mother liquor can result to rough crystals. Whatever the case may be, a mixture of powders must always be resistant to segregation by shocks or vibrations.

5.1.4. Vickers hardness of a crystal

The punch has the "diamond" shape, that is the shape of a pyramid, the tips directed to the bottom and whose base is square with diagonals d and therefore of sides $d/\sqrt{2}$.

The angle of both opposite sides of the punch is 136°. The angle made by the horizontal with these sides is (90°–136°/2) = θ.

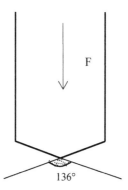

Figure 5.1. *Punch for measuring Vickers hardness*

The area of the side surfaces of the pyramid is:

$$S = \frac{d^2}{2\cos(90° - 136°/2)} = \frac{d^2}{2\sin(136°/2)}$$

The Vickers hardness D_v is a pressure equal to the quotient of the vertical force F exerted by the punch on the surface S.

$$D_v = \frac{2F\sin(136°/2)}{d^2} = 1.854\frac{F}{d^2} \qquad \text{(MPa)}$$

In order to measure the hardness of a small crystal, the force F is 25 gf and the penetration of the punch is such that the diagonals remain less than 50 μm. Aulton [AUL 77] reports and reviews the results obtained in this manner. Thus, the device described by [RID 69] was used to measure the Vickers hardness of pharmaceutical crystals with the following results obtained (MPa).

Nature of crystal	Vickers hardness
Aspirin	85
Urea	89
Potassium chloride	174
Sodium chloride	208
Sucrose	624

Table 5.2. *Vickers hardness of pharmaceutical crystals*

Thus, it is easy to make tablets with aspirin. It is not the case with powdered sugar. The softer the crystals are the easier they are to press into tablets.

5.1.5. Mohs' scale for minerals (see Appendix 1)

This scale, pronounced "moz", classifies the more common minerals according to indices varying from 0.1 to 10. Any mineral which scratches another has a greater index. Diamond has an index of 10. Mohs' scale is nothing but a scale of hardness.

5.1.6. Hiestand and Smith's indices [HIE 84]

Upon leaving the matrix after compression, the tablet can get broken. Hiestand and Smith [HIE 1984] define three indices allowing us to predict the behavior of the tablet:

– the binding index translates the persistence of real contact surfaces after decompression;

– the fragility index is measured by combining the tensile strength of a tablet with and without a hole in its center;

– the deformation index indicates the elastic recovery after elastic and plastic deformations and decompression.

The authors interpret the behavior of half a dozen excipients.

5.2. The pressing operation of powders

5.2.1. *Sintering mechanism under pressure*

The pressing of pharmaceutical tablets is nothing but a sintering operation under pressure when a set of particles in mutual contacts is subjected to a hydrostatic pressure P, and when this pressure is more than three times the value of the tensile strength of the particles. The contact points grow in the form of circles.

In other words, the resistance to plastic deformation is locally exceeded. The creeping of particles thus occurs. The rate of change of the mass density is such that:

$$\frac{1}{\rho_0}\left(\frac{d\rho}{dt}\right) \propto P_{app}^n \text{ with } n \neq 1$$

ρ: mass density of the porous sintered body ($kg.m^{-3}$)

P_{app}: applied pressure (bar)

At the points of contact, the temperature rises, which obviously promotes deformation but can also promote diffusion in such a way as to ease the stresses. The diffusion causes the rounding of pores while the plastic flow (creeping) gives the pores of the tablet a shredded shape.

The free surface of the particles grows in the pores and the spherules (or spheroids) become polyhedron. The porosity reduces and the coordination number is, let's recall, the number of contact points of each particle with its neighboring particles.

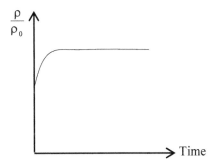

Figure 5.2. *Packing during sintering under pressure*

5.2.2. Useful additives to the pressing of tablets

We have:

1) Anti-adherent additives ("sliding") ease the flow of charge in the matrix such that the filling is uniform. The main anti-adherents are:

– talc (hydrophobic);

– starches and colloidal silica (hydrophilic). The silica can unfortunately adsorb the dissolved medicine and hinder its action.

2) Lubricants are useful to reduce friction against the tablet formation matrix.

Magnesium stearate can also act as an antiadherent on the punch and at last, it can improve the flowability of a powder. In this regard, it is therefore a "sliding" agent. Magnesium stearate is often used but it is unfortunately hydrophobic, which slows down the absorption of the medicine by gastrointestinal liquids.

Hydrophilic lubricants include polyethylene glycols (less efficient) and magnesium lauryl sulfate (as efficient as stearate).

3) Binders (or "adhesives") are useful for the maintenance of the cohesion of grains and in general (grains, tablets) to consolidate interparticle bonds.

Ethyl cellulose is a binder insoluble in water, which can slow down the assimilation of the medicine. However, a water-soluble binder like polyvinylpyrolidone gives a very viscous solution, which slows down the penetration of the solvent in the pores of a grain and especially a tablet.

We can treat a mixture of medicine powder and of a small quantity of solution of hydrophilic binder in a mixer at high speed (in order to avoid clusters).

5.2.3. Diluents

Diluents are used to increase the volume of a dose in case the amount of medicine to be absorbed is very low. The microcrystalline cellulose can be used as a diluent.

If the diluent is soluble, it will ease the dispersion of the medicine. On the other hand, if it is insoluble, it will delay the liberation and assimilation of the medicine.

The diluent can also alter the resistance of the "compacted volume" during the filling of the matrix.

5.2.4. Excipients

Excipients are substances meant to facilitate the liberation and absorption of the medicine.

1) Surfactants ensure the dampening of hydrophobic particles. In particular, they can cause the dissolution in micelles of relatively insoluble medicines.

Thus, it can recrystallize the medicine from a solution containing a surfactant. The shape (habitus) of the crystals is not modified but the drop in the fusion point suggests the presence of the surfactant on their surface. This method liberates very less amounts of surfactant in gastrointestinal liquids but have the same effect as if we were to directly add the surfactant in these liquids (the quantity would be greater and thus could be bad for the health).

If the recrystallization is done by atomization, a great proportion of the surfactant solution will go along with the crystals. It therefore seems preferable to crystallize, and dry.

2) Disintegrants centrifuge a tablet by their swelling. They develop a capillary structure favorable for the penetration of the liquid. Disintegrants other than surfactants can be divided into four categories:

– starches and their derivatives (corn, potatoes, rice). Corn starch is perfectly hydrophobic. The other starches are a little less hydrophobic.

Wheat starch swells the most but in a general sense starches swell rapidly and cause the tablet or capsule to explode. A content percentage of about 1% is sufficient;

– cellulose (microcrystalline cellulose, methylcellulose, carboxy methylcellulose, hydroxypropylcellulose, etc.) absorbs more water than starches but swells more slowly. They are usually efficient to a content capacity of the order of 2%;

– macromolecules (alginic acid, guar gum, casein formaldehyde, pectins, cation exchange resins, cross-linked polyvinylpyrrolidone, etc.) are not the best to favor the penetration of water in the tablets. In fact, they swell less. The concentration does not go above 2%.

3) Effervescent excipients are very efficient in disaggregating a tablet (bicarbonate and citric acid).

5.2.5. Description of the formation of a tablet

The dislocations exert thrusts on one another, which vary in inverse proportion with the square of the distances separating two adjacent dislocations. But, with the deformations, the dislocation density increases, the separating distances decrease and the potential energy of the dislocations network increases.

It therefore becomes progressively more difficult to introduce new dislocations as the necessary thrust increases.

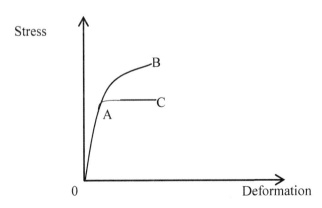

Figure 5.3. *Stress–deformation relation*

0A: elastic region

AC: rupture without hardening.

AB: rupture with hardening.

Distortion hardening during the compacting of a tablet can be seen for aspirin as the hardness of a tablet reaches 215 MPa. On the other hand, the hardness of a sodium chloride tablet is not more than that of the crystal. Meanwhile, it seems like inorganic materials often get hard due to compression just like paracetamol.

On the contrary, more voluminous molecules like sucrose, urea, lactose, microcrystalline cellulose, the crystallized maltose–dextrose mixture have a more complex structure by atomization and do not harden.

Aspirin is considered favorable for the pressing of tablets while paracetamol is much less favorable.

Hardness due to compression increases if d_p reduces by 700 to 300 μm. The hardness increases with the curvature of the surface of the tablet.

5.2.6. *Pressing operation*

The tablets are invariably prepared by compacting the medicine powder and its supplements between a punch and a matrix. The fed powder can be granulated, for example, to avoid segregation. The compaction pressure varies depending on the case from 50 to 300 MPa for tablets between 0.1 and 1 cm in diameter.

After exertion of the pressure and consolidation of the powder, there is stress relaxation. If, during this relaxation the elastic recovery is sufficiently large, these will be little or no lamination during the ejection.

The stresses due to pressing can be explained by what Huttenrauch [HUT 77] calls "activation", that is:

– a distortion of links inside the crystal, which causes the formation of amorphous regions;

– an interlocking of neighboring particles by exchange of molecules. On this aspect, we can surely speak of cold welding;

– the formation of point defects (excess or lack of molecules) or line defects (dislocation).

We have to distinguish the sliding planes (numerous) and cleavage planes, which, are trespassed by weak intermolecular links such that the crystals are easily cleavable in two parts along these planes. Thus, for sodium chloride, which crystallizes in the cubic system, the 110 planes are suitable for sliding and the 100 planes are suitable for cleavage. The cleavage leads to fractioning of the particles and the sliding can lead to the formation of dislocations. Thus, a cleaved crystal possesses on its surface alone, and in a uniform manner 10 to 10^{10} dislocations per.m^{-2}. On the other hand, a crystal subjected to pressing could have 10^{11} to 10^{16} dislocations per.m^{-2}.

In a simpler way, we can say that if the intermolecular forces are exceeded, plastic deformation along sliding planes can occur in order to restore the initial structure. But this restoration is not complete and faults are formed.

Finally, the rupture is a consequence of either cleavages or the presence of faults but it is the properties of activation which give the tablet its *cohesive nature*. Activation brings disorder to the crystal, which is shown by a small dilatation in volume (of the order of %).

When the pressing stress is sufficiently large and quick, everything happens as if the activation did not have sufficient time to occur (like a slow reaction) and rupture thus occurs.

Materials for which the hardness is limited are easily subjected to activation. Aspirin is a good example, just like lipids, while starches and cellulose are really only deformed at high pressure. There is a direct and increasing linear relationship between the necessary compacting pressure and the Vickers hardness of the dominant material in the tablet.

Tablets derived from fine crystals are the most resistant tablets. In fact, microcrystals are the hardest since during their growth they did not have the necessary time to produce faults. On the other hand, interparticle contacts are more numerous, thus, consequently, than the activation energy per unit volume.

The presence of humidity during pressing eases the transmission of energy in the powder, which increases the activation and thus the resistance of the tablets. On the other hand, the presence of lubricants dampens the disturbances in the crystalline network and reduces activation. This leads to a drop in resistance.

The tablets show two essential mechanical properties:

– Vickers hardness;

– resistance to crushing, which is equivalent to the resistance to tensile.

In summary, the manufacture of tablets is facilitated by:

– good flowability of the powder used to fill the metering tube and the matrix;

– homogeneity of the mixture;

– limiting hardness of the crystals of the powder.

As we have seen, pressing increases contact areas between particles and, consequently, ensures the cohesion of the tablet.

We add binders to the powder, which drown the particles in a matrix more or less continuous (starch, waxes). Low fusion point waxes undergo plastic deformation before the material to be agglomerated, which creates additional links between particles. The role of a binder is to attenuate the anisotropy of the stresses within the solid in contact with the wall.

Thus, for animal feeding, it is concentrated solutions of sugar (molasses) or concentrated in nitrogen (by-products of fermentation), polysaccharides, and finally fats. These products cover the particles with a very slim adherent film. Between two particles in contact the cohesion of the liquid is stronger than the adhesion energy, which is also high.

If ε_0 is the porosity of the powder at rest, and ε that of the compressed powder, Heckel's law [HEC 61] is represented by:

$$\varepsilon = \varepsilon_0 \exp(-KP) \text{ or } \ln\varepsilon = \ln\varepsilon_0 - KP$$

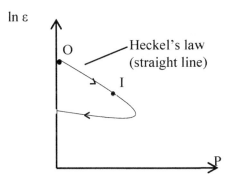

Figure 5.4. *Compression of a powder*

From the point I, the powder leaves from a divided state to a semicontinuous and cohesive state. We will find from Kawakita *et al.*'s publication [KAW 71] a complete list of compacting equations of powders.

The test of diametric break (crushing) consists in exerting a vertical thrust on a circular tablet whose flat sides are horizontal.

5.2.7. Resistance of a tablet to crushing

Resistance to crushing is related to the tensile strength as we can see by pressing a round rubber cushion, which creates a tensile stress on the entire periphery of the cushion.

More scientifically, let us consider a cylindrical tablet whose sides are horizontal. On one of the sides, we exert a thrust F. The stress corresponding to the crush is:

$$\sigma = \frac{2F}{\pi De}$$

F: distributed charge on one side (N)

D: diameter of the tablet (m)

e: thickness of the tablet (m)

σ: tensile strength (Pa).

Tensile strength is a function of the contact area between the particles and the binding force between these particles. The contact area depends on the plastic flow of the particles during pressing. It will be increased if the particles have a low resistance to this flow. The smaller the particles will be the more resistant the tablet will be as the number of mutual contacts will be increased.

Tensile stress reduces with the porosity according to Ryshkevich's law:

$$\sigma = \sigma_0 \exp(-k\varepsilon)$$

5.2.8. Distortion due to compression

We will find from Kawakita *et al.*'s article [KAW 71] 15 different expressions for the change in volume as a function of pressure. Some of these expressions are used in soil mechanics.

Equi-dimensional crystals are the most resistant to compression. On the other hand, the rods and platelets are oriented parallel to the pressing surface and, initially at least, offer less resistance to pressing.

The crystals possess three principal dimensions. Under the effect of a pressure force, the major dimension as well as the intermediate dimension are oriented parallel to the surface exerting the pressure. Thus, the rods and platelets become parallel to this surface.

5.2.9. Compression presses

The powder is supplied from a hopper in the matrix where the piston is in a low position. A scraper evacuates the excess powder. The upper piston (punch) is lowered and compresses the powder. Then, the tablet is ejected by gravity. Machines for which the matrices are disposed on horizontal discs have a production rate capable of reaching 10,000 tablets per minute.

On the other hand, Figure 5.5 shows the principle of a press for protein (or mineral) blocks for animal feeding. Chopped feed bricks are made following the same principle.

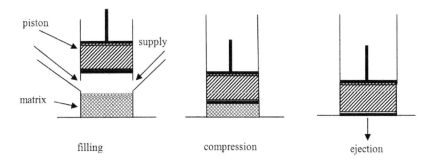

Figure 5.5. *Principle of a piston press*

The dominant parameters are:

– the flowability of the powder for the supply;

– the air release rate during compression.

The chamber (matrix) in which the piston slides presses on a horizontal closing surface, which steps aside sideways for ejection.

5.2.10. Granulation by pressing

The pressing roller forces the powder across channels. A knife then cuts through the granulated to the required length. This operation consumes a lot of energy. A premoistening (vapor) process can ensure the cohesion of the granulated.

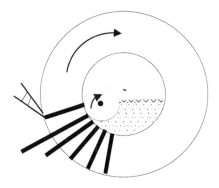

Figure 5.6. *Principle of a continuous grain*

5.2.11. Shaping of balls by compression

Two cylinders each have the half print of the balls to be achieved and are supplied in powder form by their upper part.

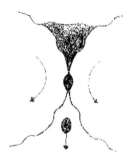

Figure 5.7. *Shaping of balls*

This device is used for the production of coke balls and also for sodium cyanide balls used in flotation.

5.2.12. Agglomeration by rapid agitation

Agglomeration by simple agitation of a powder mixture and a little amount of liquid by rapidly rotating pins or blades gives a very uniform distribution of the size of the agglomerates obtained.

5.3. The physics of rolling-granulation

5.3.1. Advantages of grains over powders

1) Grains flow more easily than powders. In fact, van der Waals forces, which give their cohesion to powders, do not act anymore as these forces are directly proportional to the diameter of the particles while the mass of these particles reduces with the cube of their diameter. The van der Waals force binding two grains is negligible compared to their weights.

2) Contrary to powders, grains do not adhere to walls and thus do not cause accumulation of the product.

3) Due to the absence of the action of van der Waals forces, granulated products do not cake during prolonged storage.

4) Granulated products do not emit dust during handling operations.

NOTE.–

In the case where the product to be granulated is a mixture of powders, we understand that mixing operation prior to granulation is difficult and must be carefully studied for each granulated product to contain every component in the right proportions.

5.3.2. States of moist divided solids

In moist aggregates, whatever they are, the liquid can be present in three different forms.

1) Pendular state: the particles are in mutual contact and, at the point of contact, the liquid is present in the form of a concave lens wetting both particles. The lenses constitute what some refer to as liquid "bridges".

2) Capillary state: the totality of interparticle voids is occupied by a liquid (the saturation is equal to 1) but the particles keep their mutual contacts.

3) Funicular state (from latin funieulus: wire. The liquid phase is continuous): this state is intermediate between the two previous states in the sense that there are always liquid bridges and at the same time there is the presence of regions entirely filled with liquid. The saturation of voids by the liquid is 0.2 or 0.3 in the funicular state.

NOTE.–

If from the capillary state, we increase yet again the liquid proportion, the particles lose their mutual contacts and we then get a simple dispersion of the particles in the liquid.

5.3.3. Capillary adhesion of a particle on a moist granulated product

Let's say (β being the half-angle which underlies the meniscus on a particle) we have:

$10 \deg < \beta < 40 \deg$

Batel [BAT 56] showed that the adhesive force of a particle of diameter d on another particle of same diameter d on the surface of a granule is:

$$F_A = \alpha \gamma d \qquad (2.2 < \alpha < 2.7)$$

(if you are a mathematics enthusiast, refer to the work of Erle *et al.* [ERL 71]. This force is very much greater than the weight of the particle of diameter *d*).

$$P = \frac{\pi d^3}{6} \rho_s g$$

ρ_s: actual mass density of the particle (kg.m^{-3})

EXAMPLE 5.1.–

$$\gamma = 0.072 \text{ N.m}^{-1} \qquad d = 10^{-6} \text{m} \qquad \rho_s = 2700 \text{ kg.m}^{-3}$$

$$g = 9.81 \text{ m.s}^{-2}$$

$$F_A = 2.2 \times 0.072 \times 10^{-6} = 0.158.10^{-6} \text{N}$$

$$P = \frac{\pi \times 10^{-18} \times 2700 \times 9.81}{6} = 0.138.10^{-13} \text{N}$$

$$0.158.10^{-6} \gg 0.138.10^{-13}$$

This calculation shows the possibility of a wet ball increasing in size by rolling on powder or simply by coming more or less in regular contact with powdered particles.

5.3.4. Steps in the formation of pellets by rolling

1) Sprinkling of water and formation of nuclei.

The water (or the binder solution) is sprayed on the mass of the granules. Random encounters take place between the liquid drops and solid particles.

By this operation, it is a matter of giving rise to agglomerates initially consisting of less elementary particles called nuclei. The result obtained depends:

– on the wettability of the particles;

– on the size of the droplets;

– on the richness of the solid mixture in fine particles and, consequently, the location of the solid mass where the drop-particle encounters take place;

– on the degree and nature of the agitation of the solid mass;

– on the flow rate of the liquid, that is the number of droplets.

Depending on the operating conditions, we have nuclei·

– more or less large;

– more or less wet.

2) Growth of the nuclei.

Whatever the material used, the device is designed for the *rolling* of granulated. In the course of this operation, different phenomena come into the picture:

– the granulated roll and, like snowballs, capture the fine, which they encounter. The increase in diameter of the pellets is directly proportional to the path, which they have travelled in the device;

– two small sized granulated (x_1 and x_2) can come together to form a pellet of diameter x. We thus have:

$$x^3 = x_1^3 + x_2^3$$

– a granulated can get ruptured.

If we increase the proportion of the liquid phase in the medium from zero, the rate at increase of the diameter of the pellets increases rapidly up to a pronounced maximum. Above this maximum, the agglomerates rapidly disintegrate and we finally obtain a slurry.

3) Characteristics of the pellets obtained.

Two factors have a positive effect on the acquisition of the required qualities. They are:

– the shocks experienced in the condition that their intensity be sufficiently moderated so as not to lead to rupture;

– a careful control of the proportion of liquid (close to maximum the growth rate).

In this case, the pellets have:

– a low porosity;

– a high resistance to crushing and abrasion, that is a high internal cohesion;

– a smooth surface;

– a quasi-spherical shape.

5.3.5. *Humidification*

It occurs:

– by spraying of water in fine droplets (d_p< 50 μm);

– or by steam jet.

Depending on the product being treated, the humidity X is contained within the following limits:

$$40\% < \frac{water}{solid} < 60\%$$

According to some authors, the amount of water required varies between 0.5 and 0.75 times, which corresponds to a plastic paste behavior for the powder.

If the supply is too dry, the load slides instead of rolling and the granulated are not formed.

For the process of granulation to begin, there should be a minimum humidity. The humidity is preferentially localized on the surface of the granules and causes plasticity thereof and, particularly, a slight deformation during impact with a particle which will help capture the latter. The particle will be retained by a liquid "bridge" (pendulum link) between granulated and particle. The plasticity of the surface pellets favors the formation and continuation of their spherical shape.

With an excessive humidity, the growth of the pellets is too quick, they are less resistant and can be destroyed when they fall under the action of their own weight. A stronger humidity joins the pellets one to the other and also to the walls. The excessive humidity can be offset by the addition of dry and fine particles.

5.3.6. Effect of drying

Drying during granulation must be such that the humidity is taken out at the same rate at which compacting takes place. In this case, the pellets will be denser. This is similar to saying that the saturation of the voids with a liquid must remain close to 100%. The granulated thus finds itself in the capillary state.

Air can be present at the center of the granulated. The global saturation is thus less than 100% and the resistance of the pellets is reduced. The fraction of air in the volume is less than 6% if the distribution is monodispersed. It can be up to 12% on a spread diameter distribution.

Drying post granulation weakens the granulated and their diameter tends to reduce by abrasion and erosion.

5.3.7. Effect of granulometry and porosity

If the supply consists of grained and fine, the porosity will reduce more drastically (compared to a monodispersed distribution) in the following conditions:

$$\frac{\text{mass of fine particles}}{\text{mass of grained particles}} = 0.25 \qquad \frac{\text{diameter of fine particles}}{\text{diameter of grained particles}} = 0.25$$

Generally, the porosity of a bulk fill of monodispersed particles is of the order of 0.35 to 0.40. It can decrease by 20% under the action of vibrations.

It happens that the agitation of pellets has the same effect as vibrations on their porosity.

The most resistant granules will be obtained by a maximum reduction in the porosity (while maintaining a porosity close to 100%).

This is how the cohesion of a compact granule is very much greater than that of loose and highly porous agglomerates. Such agglomerates have an irregular shape.

The granulometry of the pellets is such that the large ones often have three times the diameter of the small ones.

5.3.8. Kinetics of continuous granulation

The growth of the aggregates by the snowball effect is compared to the growth of a crystal from a mother solution.

For a perfectly mixed (homogenous continuous) crystallizer, we define N as being the total number of crystals contained in $1 \, m^3$ of the product inside the granulator. There are heterogeneities inside of such a device, but we will neglect them and assume that their average internal granulometry is equal to that of the evacuated product. This granulometry is described by a population density n(x) such that:

$$n(x) = N \frac{dP}{dx}$$

where P is the fraction of underflow across a screen of opening x.

n (x) is measured in m^{-4}.

For such a crysallizer, we show that the number of crystals formed around the size x is, for $1 \, m^3$ of product and during the time $\Delta\tau$.

$$[n(x - \Delta x) - n(x)]\Delta x = -\Delta n(x)\Delta x = -\Delta n(x)G\Delta\tau$$

where G is the rate of enlargement of the crystals: $G = \Delta x/\Delta\tau$.

If V_p is the apparent volume occupied by the product in the device, the number of "crystals" (for us, the number of agglomerates) formed in the section of width Δx approximately equal to x is:

$$-V_p \Delta n(x) G \Delta \tau$$

If Q_0 is the apparent (that is with voids included), flow rate of the extracted product, the number of agglomerates formed is:

$$Q_0 n(x) \Delta x \Delta \tau$$

Equating these two expressions, we obtain:

$$-G \frac{dn}{dx} = \frac{Q_0}{V_p} n(x) = \frac{n(x)}{\tau} \qquad [5.1]$$

τ is the average dwelling time of the product in the granulator.

The agglomerates grow in size and turn into pellets by the snowball effect. The increase in diameter is proportional to the path travelled by the agglomerates and we have seen that the latter move a lot quicker as their sizes increase. We therefore assume that for G the simplest possible law, that is a proportionality between G and the diameter x of the agglomerates:

$$G = gx$$

However, the previous equation is not sufficient as the snowball effect is not the only one coming into picture:

– an agglomerate can break;

– a new agglomerate can result from the reunion (coalescence) of two smaller agglomerates.

Several researchers have sought approximate analytic expressions for the appearance rates by coalescence and for disappearance by rupture, but to no avail. We shall proceed differently and add to the first term of equation [5.1] a velocity C(x), which, if it is positive, will correspond to the dominant appearance of grains by coalescence and, if it is negative, will correspond to the disappearance of the grains by rupture.

The agglomeration equation is therefore written as:

$$-\frac{dn}{dx}gx + C(x) = \frac{n}{\tau}$$

5.3.9. Expression for the rupture-coalescence velocity

The product of a granulator is relatively tightened in granulometry and we can express the fraction of the cumulated flow across a sieve of opening x by an expression of the form:

$$P = \frac{1}{1 + \left(\dfrac{x_{50}}{x}\right)^{\lambda}} = \frac{x^{\lambda}}{x^{\lambda} + x_{50}^{\lambda}} \qquad (\lambda > 1)$$

It is an equation in two unknowns, λ and x_{50}.

The differential concentration is therefore:

$$n(x) = N\frac{dP}{dx} = \frac{N\lambda x_{50}^{\lambda} x^{\lambda-1}}{\left(x^{\lambda} + x_{50}^{\lambda}\right)^2}$$

N: apparent number of agglomerates per m^3 of the product in the device.

Similarly:

$$\frac{dn(x)}{dx} = N\frac{x_{50}^{\lambda} x^{\lambda-2}\lambda}{\left(x^{\lambda} + x_{50}^{\lambda}\right)^3}\left[x_{50}^{\lambda}(\lambda-1) - x^{\lambda}(\lambda+1)\right]$$

From the agglomeration equation:

$$C(x) = \frac{1}{\tau}\left(n + g\tau x\frac{dn}{dx}\right)$$

that is:

$$C(x) = \frac{N\lambda x_{50}^{\lambda} x^{\lambda-1}}{\left(x^{\lambda} + x_{50}^{\lambda}\right)^3}\left[x_{50}^{\lambda}\left(1 + g\tau(\lambda-1)\right) - x\lambda\left(g\tau(\lambda-1)-1\right)\right]$$

The previous equations allow us to find:

– the mode of the distribution, that is the value x_m of x for which $dn/dx = 0$.

– the critical size x_c for which the ruptures balance out the coalescences, that is for which $C(x) = 0$.

EXAMPLE 5.2.–

In order to obtain an order of magnitude for g, let us consider an isolated pellet with 3 mm diameter obtained from fine products of 100 μm size, and let us assume that this result was obtained within an hour.

We have:

$$\frac{dx}{d\tau} = gx$$

that is:

$$Ln\frac{x}{x_0} = g\tau$$

hence:

$$g\tau = Ln\left(\frac{3.10^{-3}}{10^{-4}}\right) = 3.4$$

so:

$$g = 9.4.10^{-4}\, s^{-1}$$

If the granulometry is tightened up: $\lambda = 5$

hence:

$$\frac{x_m}{x_{50}} = \left(\frac{\lambda-1}{\lambda+1}\right)^{1/\lambda} = 0.92$$

$$\frac{x_c}{x_{50}} = \left[\frac{1+g\tau(\lambda-1)}{g\tau(\lambda+1)-1}\right]^{1/\lambda} = \left[\frac{1+13.6}{20.4-1}\right]^{0.2} = 0.95$$

NOTE.–

1) Experience allows us to reach values of τ, x_{50}, and λ and if, in addition, we have approximate expressions for $C(x)$ and $n(x)$, the agglomeration equation will give the value of g or, more generally $g(x)$.

2) We have made use of an equation in two unknowns for the accumulated overflow on a sieve, but it is possible to use more complicated expressions, for example:

$$R = \left(1 + \alpha x^q\right) \exp\left(-px\right)$$

This equation has been used by Murthy *et al.* [MUR 82]. It is an equation with three unknowns.

5.4. Granulating equipment

5.4.1. *Drum*

It is followed by a screen, which allows the recycling of agglomerates of insufficient sizes. The recycled flow rate can represent 1–4 times the production.

Circular weirs at the entrance and exit enable us to determine a satisfactory dwelling time.

The spraying of liquid is done close to the entrance and on the product in cascade and not on the walls. Sometimes the liquid is sprayed on a product curtain falling from lifter blades.

In order to obtain a flow of the product in cascade but with shocks of moderate intensity, the rotation speed varies from 0.3 to 0.5 times the critical speed $N_{c\text{-}}$:

$$N_c = \frac{\pi}{\sqrt{gD}} = \frac{42,3}{\sqrt{D}} \qquad (\text{rev./minute})$$

where D is the diameter of the drum, in meters.

To a given fraction of the critical speed corresponds a well-defined cascade regime of pellets and especially a specific value of the energy lost during impact between pellets and agglomerates.

The granulation path travelled per unit time is proportional to ND and, in a constant regime, proportional to:

$$kN_cD = k \times 42.5\sqrt{D}$$

The axial progression is proportional to the product of the previous path by the slope of the drum and by the cross-sectional area of the drum (at a constant filling ratio).

The production P is therefore:

$$P \propto D^2 \times \sqrt{D} = D^{2.5}$$

5.4.2. Operating parameters of the drum

1) Rotational speed:

Ø drum (m)	Peripheral speed (m.s^{-1})
0.45	0.50
2.5	0.80

Table 5.3. *Order of magnitude of the rotational speed*

The rotational speed must be such that:

$$0.30N_{critical} < N < 0.5N_{critical}$$

The size distributor gets wider of the pellets if the speed reduces and the granulation period increases as it is a direct function of the number of turns made by the drum.

2) Filling of the drum:

The load should represent in the order of 5–10% of the volume of the drum. If the load increases, the granulation is faster but there can be cascade flow (and not rolling), which may break the pellets.

3) Granulation period:

The granulation period is directly proportional to the desired size for the granules and inversely proportional to the size of the particles.

5.4.3. Plate

It combines the granulation with the sieving in a single operation as, on the plate, there is segregation and the fine accumulate at the bottom while the pellets float and are discharged by the peripheral weir of height H (in general H/D < 0.2).

The thrusts experienced by the pellets are the same as those in the drum:

$$N = 0.75N^c$$

$$N_c = \frac{42.3}{\sqrt{D}} \sin \alpha$$

α: plate angle on the horizontal plane

α varies from 45 to 70°. The diameter of the pellets reduces with the angle α (that is with the dwelling time).

D varies from 1 to 6 m.

The filling is of magnitude ranging from 0.15 to 0.20 ton.m^{-2}.

The minimum flowrate is:

$$W = 1.5D^2 \qquad (t/h)$$

(D in meters), the flow rate can be up to 100 ton.h^{-1} in the most favorable situations.

This equipment is compact.

A scraper sweeps the bottom surface and the weir. A layer of product on the metallic wall avoids slipping and facilitates the rolling of the pellets.

The minimum power consumed is:

$$P = 1.2D^2 \quad (kW)$$

D is expressed in meters.

The dwelling time increases with:

– the height of the weir;

– the horizontality of the plate.

The position of the supply has an effect on the size of the pellets. If we give the supply:

– at the lower portion of the plate, there the pellets overflow and are abundant on the surface, the fine will be firmly positioned on the pellets at this point and the size of the product obtained will be maximum.

– on the contrary, if we give the supply at the upper portion, the fine will have the time to agglomerate with each other and the size obtained will be minimum.

Note that we can slightly increase the production by increasing the rotational speed and the slope of the plate, but the pellets will be smaller.

5.4.4. Comparison between drum and plate

The plate has the following advantages:

– classification is efficient (the granules are on the surface and close to the exit) and that allows the spraying of the supply at the right place;

– if the supply occurs in the supersaturated state (urea, ammonium nitrate), cooling by the surrounding allows for a lower recycling ratio of the fine than for the drum;

– visual surveillance is easy.

The drum has the following disadvantages:

– smoke and dust are difficult to perceive;

– in the case of fertilizers, ammoniation is better achieved in a drum.

5.4.5. Dynamic elements of divided solids

Partially fill a rotating drum with solid grains. Due to the rotation of the drum, the solid is led by the wall and this solid then falls back like an inclined plane. Movement of the solid is then created as it is shown in the Figure 5.8.

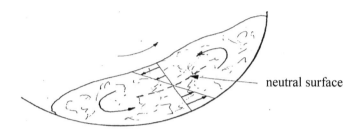

neutral surface

Figure 5.8. *Movement of solid in a rotating drum*

A surface, which we will call as neutral surface, separates the ascending solid from the descending solid. The movement of the solid cannot be calculated simply as divided solids (D.S.) do not behave like fluids. In particular, the shear stress as well as the normal stress are given by an equation of the form:

$$\tau = \tau_0\left(\beta\right) + \mu\left(\beta\right)\rho_s d_g^2 \left(\frac{dU_x}{dy}\right)^2$$

β: compactness (1's complement of the porosity)

ρ_s: actual density of the solid (kg.m^{-3})

d_g: diameter of the grains (m)

The velocity profile is therefore far from being linear and can have different forms. It is important to notice that a segregation acts between the fine and the grained and that these latter cover more distance as they rotate outside.

A second type of segregation arises. In a continuous granular drum, the average size of the grains increases from the bottom to the surface.

The movement on a plate is similar to that of an inclined drum, with the slight difference that the bottom of the drum is filled with small agglomerates.

NOTE.–

Holley [HOL 83] gives useful information on the working of the drum and the plate. In addition, he proposes a mixer said to be "turbulent" constituted of a shaft with needles and rotating in a horizontal cylinder. This mixer is useful for the addition of 3–5% of water to a D.S. in such a way as to avoid dusts.

5.5. Resistance of the granular

5.5.1. Deformation due to tensile strain [SCH 75a, SCH 75b]

The tensile strength is critical for the resistance to crushing (see section 5.2.7).

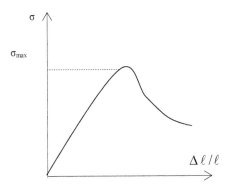

Figure 5.9. Tensile test on a test tube

The decreasing part of the curve $\sigma = f(\Delta \ell / \ell)$ represents plastic sliding of the particles one on the other. The maximum stress σ_{max} decreases with the length of the test tube as a long test tube contains more zones of weakness than a short test tube.

5.5.2. Effect of saturation on σ_{max}

When the saturation varies, the stress σ_{max} reaches its maximum value in the capillary state. It varies linearly with saturation in the funicular state [PIE 69].

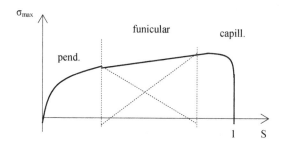

Figure 5.10. *Effect of saturation*

5.5.3. Cohesion of pellets

In order to meet the previous requirements, the pellets must resist crushing and mutual abrasion and, for that to happen, they must possess a certain cohesion.

This cohesion increases with:

– the number of bridges linking each particle to its neighbors and

– the solidity of these bridges.

The number of bridges linked to a particle is equal to its coordination number which, in practice, is of the order of 6. This index increases with the compactness of the pellet, that is when its porosity reduces. The shocks received by the pellet during granulation contribute to the removal of air pockets, such that the porosity reaches the desirable value of 30%. This value is less than that of a bulk stacking of spheres (approximately 40%) and can be explained by the fact that the granulometry of the particles is spread and the fine occupy the free space between the granules.

The bridges linking the elementary particles are especially more per unit volume of agglomerated material as:

– the particles are fine (their size must be less than 100 μm);

– they have a shape that looks less spherical (dendrite, plates).

Both conditions are expressed by a particular high surface area, which can be measured by making air pass through a bed of particles in laminar regime.

The nature (and therefore the solidity) of the interparticle bridges varies according to the treated product:

– fertilizers such as urea or ammonium nitrate are sprayed and the interparticle bonds correspond to a mere solidification by crystallization;

– a product like limestone can be agglomerated by spraying of water. The water is adsorbed on the surface of the particles and the bridges are made of molecular layers of water. These molecules, by hydrogen bonding, ensure the solidity of interparticle bridges. But the cohesion of the pellets thus obtained is weak;

– in order to agglomerate an iron ore intended to charge blast furnaces, water alone is not enough and we have to incorporate a binder to it. Bentonite swells by absorbing water and develops hydrogen bonds with ore particles. If we're dealing with a soluble silicate, it will vitrify at high temperatures and cement the particles between them. In fact, in some cases, the binder used is purely and simply cement.

Food products and washing powders don't appear as pellets but as granular products in different forms. In the presence of water, particles dissolve partially on the surface then, by moderate heating, there is evaporation of water and formation of solid bridges between the particles by crystallization. The food products do not necessarily crystallize and there is therefore the formation of a glassy solid.

5.5.4. Tensile strength of capillary origin

The hydraulic radius of the average pore is [DUR 99]:

$$R_0 = \frac{2\varepsilon}{(1-\varepsilon)s}$$

ε: porosity

s: volume area of the solid (m^{-1})

$$s = 6 \sum_i \frac{m_i}{d_i} \qquad\qquad \left(s = \frac{6}{d} \right)$$

m_i: mass fraction of the particles of size d_i

The capillary depression is (law of Laplace).

$$\Delta P_c = \frac{2\gamma}{R_0} = \frac{\gamma(1-\varepsilon)s}{\varepsilon}$$

The tensile strength only acts in regions occupied by the liquid that is the fraction ε of the area:

$$\sigma = \varepsilon\Delta P_c = \gamma(1-\varepsilon)s$$

γ: surface tension ($N.m^{-1}$)

EXAMPLE 5.3.–

$$d_s = 10^{-6}\, m \qquad \gamma = 0.072\, \mu.m^{-1} \qquad \varepsilon = 0.36$$

$$\sigma = 0.072(1 - 0.36) \times 6/10^{-6}$$

$$\sigma = 4.32.10^5\, Pa = 4.32\, bar$$

In fact [RUM 58], for the wet granulated (pellets) (green) of ore:

0.5 bar $< \sigma <$ 5 bar

By drying the pellets, their resistance reduces:

0.1 bar $< \sigma <$ 0.5 bar

5.5.5. *Van der Waals forces*

The force of attraction between two spheres is:

$$F_A = \frac{Ad}{24h^2}$$

A: Hamaker's constant $(10^{-19}$ J)

d: diameter of particles (m)

h: distance between the surfaces of both particles

EXAMPLE 5.4.–

$$A = 10^{-19} \text{ J} \qquad d = 10^{-6} \text{ m} \qquad h = 10 \text{ nm} = 10^{-8} \text{ m}$$

$$F_A = \frac{10^{-19} \times 10^{-6}}{24 \times 10^{-16}} = 4.16.10^{-11} \text{N}$$

Let us now assume that the attraction occurs between the plane protuberances of roughness. The distance between plane surfaces is ten times weaker, say 10^{-9} m. The fraction of surface occupied by the plane projections is 5% of the surface of a sphere considering the fact that these plane surfaces are in direct contact. The concerned surface is therefore:

$$S = \pi d^2 \times 0.05 = \pi \times 10^{-12} \times 0.05 = 1.57.10^{-13} \text{ m}^2$$

The force of attraction becomes, between plane protuberances:

$$F_A = \frac{AS}{6\pi h^3} \qquad \text{[RUM 58]}$$

Hence:

$$F_A = \frac{10^{-19} \times 1.57.10^{-13}}{6\pi \times 10^{-27}} = 8.32.10^{-7} \text{N}$$

5.5.6. *Tensile strength exerted by van der Waals forces*

The number of particles per unit cross-sectional area is:

$$n_p = \frac{1-\varepsilon}{\pi d_s^2/4}$$

ε: porosity

d_s is an average diameter corresponding to an average cross-sectional sectional area of the particles.

$$\left(\Sigma m_i\right)\left(\frac{\pi d_s^2}{4}\right) = \Sigma m_i \frac{\pi d_i^2}{4}$$

The number of interparticle contacts per unit cross-sectional area is:

$$n_c = n_p \frac{k}{2} = \frac{(1-\varepsilon)k}{\left(\pi d_s^2/4\right)2}$$

k: number of contacts per particle (coordination number)

The coefficient ½ comes from the fact that there exists bonding for two particles.

The resistance per unit cross-sectional are is given by:

$$\sigma = n_c F_A = \frac{(1-\varepsilon)k F_A}{\pi d_s^2}$$

In a porous medium, based on the results obtained by Smith *et al.* [SMI 29] we can accurately say that:

$$k\varepsilon \# \pi \qquad (0.25 < \varepsilon < 0.5)$$

Finally:

$$\sigma = \frac{(\lambda - \varepsilon)}{\varepsilon d^2} F_A$$

EXAMPLE 5.5.–

$\epsilon = 0.36$ capillary force (section 5.5.4) $F_A = 0.158.10^{-6}$ N

$d = 10^{-6}$ m   van der Waals force (section 5.5.5) $F_A = 8.32.10^{-7}$ N

Hence the following expressions for the tensile strength:

$$\sigma = \frac{(\lambda - 0.36)}{0.36 \times 10^{-12}} F_A = 1.78.10^{12} F_A$$

Capillary action $1.78.10^{12} \times 0.158.10^{-6} = 2.81.10^5 \, Pa = 2.81 \, bar$

Van der Waals $1.78.10^{12} \times 8.32.10^{-7} = 14.80.10^5 \, Pa = 14.8 \, bar$

5.5.7. Formation of a solid bridge between particles

We can consolidate a granulated obtained by rolling by using not pure water but a salt solution. This way, if subsequent drying is intended the salt crystallizes between the particles and thus forms bonds in the solid state.

We shall use the ability of porosity to remain unchanged across a cross-sectional area or a volume [DUR 99].

Let ϵ_p be the volume of the crystals bridges in 1 m^3 of granule. The corresponding mass is:

$$\epsilon_p \rho_s = (1 - \epsilon) \rho_{GR} Xx$$

ρ_s and ρ_{GR}: mass density of the crystals and the granulate

ϵ: porosity

X: moisture content of the product: mass of solution/mass of solid

x: mass fraction of the solute in the solution

Here we assume that the humidity appears in pendulous form that is that the solution creates "bridges" between the particles.

The solid bridges obtained after drying are not necessarily single crystals. We will admit that, averagely, their tensile strength is of the order of 50 bar.

The tensile strength of the granulated becomes:

$$F_{tra} = \varepsilon_p \times 50 = \frac{50(1-\varepsilon)\rho_{GR} Xx}{\rho_s}$$

EXAMPLE 5.6.–

$\varepsilon = 0.35$ $x = 0.29$ $X = 0.07$

$\rho_s = 2\ 700\ kg.m^{-3}$ $\rho_{GR} = 1\ 500\ kg.m^{-3}$

$$\varepsilon_p = \frac{(1-0.35)\times 1500 \times 0.07 \times 0.29}{2\ 700}$$

$\varepsilon_p = 0.0073$

$F_{tra} = 50 \times 0.0073 = 0.36\ bar$

We notice that the formation of salt bridges is not critical compared to capillary action or van der Waals forces.

5.5.8. Effect of granulometric distribution [LIN 73]

A spread distribution (consisting of fine products) produces granules that are less porous and highly resistant to crushing. The agglomerates grow in size by coalescence.

With a narrow distribution (without fine particles), the porosity increases and the resistance of the granules reduces. Growth is done by crushing (due to collisions) and coating of granules by successive layers.

NOTE.–

The diameter of the agglomerates follows according to [LIN 73] a distribution that is similar to the lognormal law.

5.5.9. Conclusion

We notice that the tensile strengths are close when they are determined:

– from pores (see section 5.5.4), say 4.32 bar,

– from contact points (see section 5.5.6), say 2.81 bar.

On the other hand, the resistance due to van der Waals forces (see section 5.5.6) is clearly greater (14.8 bar). This resistance is obtained not by rolling in a rotating drum or a granulated plate but by a pressure that draws the particles closer.

In fact, agglomeration presses can develop pressures of the order of a thousand bars. The particles thus pressed warm up and molecular diffusion can take place in both directions across each contact microsurface between the particles.

Thus, a couple of particles draws closer to a unique particle defined by a certain tensile breaking strength and by a modulus of elasticity. That is how:

– for lignite briquettes   100 bar < R < 300 bar

– for ore briquettes   R > 60 bar

NOTE.–

Juhasz [JUH 85] investigated the work done by compaction from the porosity of the D.S. and the volume area of the particles.

Couroyer *et al.* [COU 00] investigated the attrition of particles during the compression of a powder.

6

Mechanics and Thermics
of Gaseous Fluidized Beds

6.1. Mechanics of gaseous fluidized beds

6.1.1. Mechanism of fluidization

Let us consider a set of particles placed in a vertical cylinder of which the bottom surface is made up of a perforated plate. For this ideal experiment, we will assume that the plate, called the distributor, has holes of diameter less than that of the particles. The role of the distributor is to uniformly distribute the gaseous flux on the cross-sectional area of the cylinder.

When in the cylinder and for the first time, we increase the gaseous flow rate from 0, we notice that the drop in pressure ΔP across the bed of particles occurs according to the curve (a).

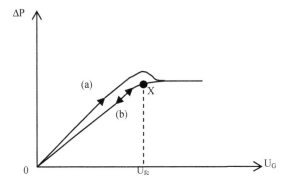

Figure 6.1. *Drop in pressure as a function of the gas velocity*

The curve (a) possesses a maximum, which is justified by the energy that was necessary to free each particle from its neighboring particles. In fact, in a fixed bed, the particles are trapped in between their neighboring particles. This step corresponds to an increase in the porosity of the bed starting with the upper layers. This step is then followed by a drop in ΔP after its maximum.

If we now reduce the gaseous flow rate, the drop in pressure ΔP follows the curve (b). The particles gently position themselves on each other while maintaining the porosity ε_{fc} higher than the porosity ε_0, which the initial bed of particles possessed when more or less compressed.

Now if we again increase the gaseous flow rate, the drop in pressure ΔP increases, but this time following the straight line (b) with a porosity ε_{fc}, which we refer to as the incipient fluidization porosity. The straight line OX represents Ergun's law for this porosity ε_{fc}. Above the point X, the increase in the flow rate U_G no longer causes ΔP to vary and it is instead the porosity that starts rising. We say that the bed is expanding.

Thus, notice that the drop in pressure ΔP remains constant when the velocity of the gas in an empty bed goes beyond the value U_{fc} at the incipient fluidization. The reason is that the gas supports the weight of the solid particles, which is itself constant.

$$\Delta P = (\rho_S - \rho_G) g (1 - \varepsilon) H$$

$$\Delta P = \text{const.}$$

H: height (m) of the fluidized bed corresponding to the porosity of the bed.

The mass density ρ_S of the solid was adjusted by buoyancy thus the subtraction of ρ_G. But (except when the fluid is a liquid) this modification is generally negligible for gases.

NOTE.–

Anderson and Jackson [AND 67a, AND 67b] suggested an interpretation of the behavior of fluidized beds based on fluid mechanics. But practical applications are absent.

Subsequently, Pritchett *et al.* [PRI 78] developed an efficient numerical method for the simulation of a gaseous fluidized bed, meanwhile Donsi *et al.* [DON 84] developed a high-voltage discharge probe for the study of bubbles.

6.1.2. *Formation of bubbles*

When we continue increasing the gaseous flowrate above the incipient fluidization velocity, the bed expands but remains homogenous. When we reach the incipient bubbling velocity, we notice:

– the formation of bubbles in the fluidized bed at a distance of the order of a few centimeters above the distributor [DUR 02a];

– the subsidence of the homogenous fluidized phase whose porosity ε_h regains the value ε_{fc};

– that the overall porosity ε_{gl} of the fluidized bed stabilizes as from the point S.

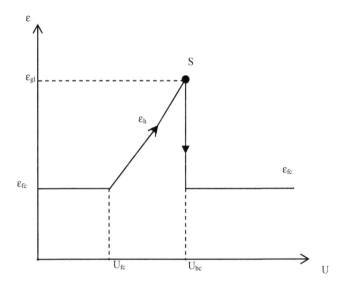

Figure 6.2. *Porosities ε_h and ε_{gl} as a function of the gas velocity*

6.1.3. *Distributor design*

The drop in gas pressure across the distributor must be greater than or equal to 0.25 times the drop in pressure across the bed for the flow rate to be uniform across the cross-sectional area.

The injectors constituting the distributor can be:

1) single nozzles spitting vertically upward. The drop in pressure corresponding to the brutal contraction at the entrance of the nozzle is:

$$\Delta P_{CO} = \rho_G \left(\frac{3}{4} V_{tu}^2 - \frac{U^2}{2} \right)$$

V_{tu}: velocity at the neck of the nozzle (m.s^{-1})

U: velocity of the gas in an empty bed (m.s^{-1})

ρ_G: mass density of the gas (kg.m^{-3});

2) nozzles debiting horizontally as they have an end containing two or four openings. The drop in pressure corresponding to an abrupt change in direction at right angle is:

$$\Delta P_L = \frac{\rho_G V^2}{2}$$

It is this last device that is preferable in order to avoid the solid from passing through distributor when we stop the gas flow rate.

6.1.4. *Classification of particles [GEL 73]*

The particles can be classified into four groups (see Figure 6.3):

1) Group A: the particles are rather fine. The bubbles activate the mixture of particles as their velocity is greater than that of the interstitial gas. The exchange of gas–solid matter is important, hence the use of these particles as catalysts. Bubbling starts after a considerable expansion of the homogenous bed. In other words:

$$U_{bc} / U_{fc} \gg 1$$

2) Group B: the particles can reach a size of 1 mm and grains of sand are the most characteristic example. Bubbling starts almost immediately after the start of fluidization;

3) Group D (D as in "dense"): the size of the particles is greater than, 500 μm and have a high mass density. The mixture of the solid is weak as the velocity of most of the bubbles does not go beyond that of the interstitial gas. A powder belongs to group D if $u_b < U_{fc} / \varepsilon_{fc}$. As a result of the agitation due to the bubbles, the sticky particles can be treated;

4) Group C (C as in "cohesive"): the particles are very fine and of low density. Thus, the interparticle forces of attraction are manifested by the formation of agglomerates that, in principle, we can attenuate through the following means:

– vibrations;

– addition of fine silica (d # 1 μm);

– humidification of the gas to eliminate the electrostatic charges.

Finally, Group C particles are difficult to fluidize.

We will find in Baeyens and Geldart's book [BAY 73], a detailed description of the different properties of the groups of particles.

Figure 6.3 must especially be considered as having a qualitative value. In fact, all the trials have been carried out with air under normal conditions (1 bar abs. and 20°C).

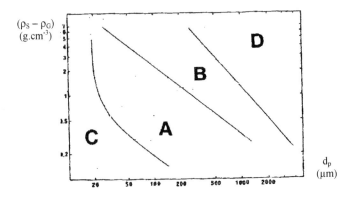

Figure 6.3. *The four particle groups [BAY 73]*

6.1.5. *Coefficient of geometric sphericity of the particles*

For a given particle, the coefficient of geometric sphericity ψ_0 is defined as:

$$\psi_0 = \frac{\text{Surface area of the sphere with the same volume as the particle}}{\text{Surface area of the particle}}$$

If S_p and V_p are the surface area and the volume of a particle, we can say that:

$$S_p = \pi d_S^2 \quad \text{and} \quad V_p = \frac{\pi d_V^3}{6}$$

S_p can be found by applying Ergun's formula in a laminar regime (see [DUR 99]) and V_p can be found by weighing a known number of particles and using the actual mass density of the particles.

By definition of d_{SV}, we can write:

$$\sigma = \frac{6}{d_{SV}} = \frac{S_p}{V_p} = \frac{6\pi d_S^2}{\pi d_V^3} \quad \text{hence} \quad d_{SV} = \frac{d_V^3}{d_S^2} \qquad [6.1]$$

Hence:

$$\psi_0 = \frac{\pi d_V^2}{\pi d_S^2} = \frac{d_V^2}{d_S^2} \qquad [6.2]$$

By eliminating d_S^2 from [6.1] and [6.2], we obtain for the coefficient of sphericity:

$$\psi_0 = \frac{d_{SV}}{d_V} < 1 \qquad [6.3]$$

For other polyhedrons, we shall refer to Table 6.1, keeping in mind that the non-sphericity index is the reciprocal of the coefficient of sphericity.

Tetrahedron		0.67
Octahedron		0.83
Ellipsoids	1 : 1 : 2	0.93
	1 : 2 : 2	0.92
	1 : 2 : 4	0.79
	1 : 1 : 4	0.78
	1 : 4 : 4	0.70

Table 6.1. *Geometric sphericities*

6.1.6. *Practical sphericity*

Recall that each particle possesses three dimensions corresponding to a trisectangular system of coordinates:

– the minor dimension;

– the intermediate dimension;

– the major dimension.

A particle can pass through a sieve hole if the diameter of this hole is greater than the intermediate and minor dimensions of the particle. The major dimension can be far greater than the diameter of the hole and the greater this dimension is, the greater the ratio:

$$\tau = \frac{d_V}{d_O} > 1$$

d_O: screen opening

Thus, Abrahamsen and Geldart [ABR 80] measured for quartz:

$\tau \# 1,13$

The important physical quantity is d_{SV} and, from [6.3]:

$$d_{SV} = \psi_O d_V = \psi_O \tau d_O = \psi d_O$$

ψ: practical sphericity

Products	ψ
Crushed coal	0.75
Mica	0.28
Angular sand	0.70
Eroded sand	0.95
Common salt	0.84
Crushed glass	0.65

Table 6.2. *Sphericity of common divided solids*

6.1.7. Average volume area

For the granulometric section of index i representing the mass fraction m_i of the divided solid taken into consideration, the partial volume area is:

$$\Delta\sigma_i = \frac{6m_i}{d_{SVi}}$$

The average volume area is:

$$\sigma = \sum_i \frac{6m_i}{\psi d_{oi}}$$

6.2. Flow thresholds

6.2.1. Porosity at the incipient fluidization

The porosity of a divided solid is the fraction of its volume corresponding to holes. In what follows, we will build on the works of Brown *et al.* [BRO 50] and of Leva [LEV 57].

The packed porosity ε_0 is obtained after a few taps on the test tube containing the divided solid. This porosity can be reproduced easily. It depends on the size of the particles and their coefficient of sphericity. We will write:

$$d_p > 500 \ \mu m \qquad \varepsilon_o = \exp\left(0.06 - 1.2\psi\right)$$

$$d_p < 500 \ \mu m \qquad \varepsilon_o = \left(\frac{500}{d_p}\right)^{0.122} \exp\left(0.06 - 1.2\psi\right)$$

ψ: sphericity index

d_p: diameter of particles: μm

In these expressions, the work of Brown et al. [BRO 50] used.

When the granulometry is not uniform but spread, the fine grains occupy the holes left between the larger grains such that the porosity is reduced in a proportion of the order of 5–20% of the value of ε_0.

The incipient fluidization porosity ε_{fc} is most certainly close (to within 1%) to the unpacked porosity and there exists a correspondence between the packed porosity ε_0 and the incipient fluidization porosity.

$$\varepsilon_{fc} = \varepsilon_o \left(1 + 0.18\psi^{0.5}\right)$$

EXAMPLE 6.1.–

$d_p = 100 \ \mu m \qquad \psi = 0.70$ (angular sable)

$$\varepsilon_o = 5^{0.122} \exp\left(0.06 - 1.2 \times 0.70\right) = 0.55$$

$$\varepsilon_{fc} = 0.55 \times \left(1 + 0.18 \times 0.70^{0.5}\right)$$

$\varepsilon_{fc} = 0.63$

6.2.2. Incipient fluidization velocity

At the start of fluidization, and for the particles of sizes less than 100 μm, the velocity of the gas in an empty bed is given by the equation [BAY 73]:

$$U_{fc} = \frac{0.0009 \left[g\left(\rho_S - \rho_G\right)\right]^{0.934} d_p^{1.8}}{\mu^{0.87} \rho_G^{0.066}}$$

U_{fc}: incipient fluidization velocity (m.s^{-1})

g: acceleration due to gravity: (9.81 m.s^{-2})

ρ_S and ρ_G: mass densities of the solid and gas (kg.m^{-3})

d_p: diameter of the particles: m

μ: viscosity of the gas: Pa.s

For the diameters of particles of sizes greater than 100 μm, the uncertainty in the porosity is smaller and some authors used Ergun's equation:

– by substituting the coefficient 150 by the coefficient 180 of von Carman;

– by assigning a value of 0.4 to ε_{fc}.

Ergun's equation then becomes:

$$\frac{1}{(1-\varepsilon)}\frac{\Delta P}{L} = 1687\frac{U_\mu}{d_{sv}^2} + 27.3\frac{\rho_G U^2}{d_{sv}} = (\rho_S - \rho_G)g$$

By taking:

$$Re_{fc} = \frac{\rho_G U_{fc} d_p}{\mu} \quad \text{and} \quad Ar = \frac{\rho_G d_p^3 (\rho_S - \rho_G)g}{\mu^2}$$

we get, for the Reynolds number, a quadratic equation whose physical root is:

$$\varepsilon_{fc} = \frac{\mu}{\rho_G d_p}\left[(954.8 + 0.037 Ar)^{1/2} - 30.9\right]$$

EXAMPLE 6.2.–

Kerosene cracking catalyst (from the ancient Greek word *keros*, which means candle as it is in fact kerosene):

$$\rho_S = 1\ 100 \text{ kg.m}^{-3} \qquad \mu = 0.008.10^{-3} \text{ Pa.s}$$

$$\rho_G = 2.6 \text{ kg.m}^{-3} \qquad d_p = 60.10^{-6} \text{ m}$$

$$U_{fc} = \frac{9.10^{-4} \times \left[(1100 - 2.6) \times 9.81\right]^{0.934} \times \left(60.10^{-6}\right)^{1.8}}{\left(0.008.10^{-3}\right)^{0.87} \times 2.6^{0.066}}$$

$$U_{fc} = 0.003366 \text{ m.s}^{-1}$$

EXAMPLE 6.3.–

Drying of a plant:

$$\rho_S = 900 \text{ kg.m}^{-3} \qquad \mu = 0.020.10^{-3} \text{ Pa.s}$$

$$\rho_G = 1.2 \text{ kg.m}^{-3} \qquad d_p = 10^{-3} \text{ m}$$

The Archimedes number is:

$$Ar = \frac{1.20 \times 10^{-9} \times 900 \times 9.81}{\left(2.10^{-5}\right)^2} = 6490$$

$$U_{fc} = \frac{2.10^{-5}}{1.2 \times 10^{-3}} \left[(954.8 + 0.037 \times 6490)^{1/2} - 30.9 \right]$$

$$U_{fc} = 0.061 \text{ m.s}^{-1}$$

6.2.3. Incipient bubbling velocity

During incipient bubbling, bubbles are formed approximately 5–10 cm from the grid. The corresponding velocity in an empty bed is given by Abrahamsen and Geldart [ABR 80]:

$$U_{bc} = 2.07 \frac{d_p \rho_G^{0.06}}{\mu^{0.347}} \exp\left(0.716 \, F\right)$$

U_{bc}: velocity in an empty bed during incipient bubbling: m.s^{-1}

d_p: diameter of particles: m

ρ_G: mass density of the gas: $kg.m^{-3}$

μ: viscosity of the gas: Pa.s

F: mass fraction of the particles of size less than 45 μm

EXAMPLE 6.4.–

Cracking catalyst:

$$\rho_G = 2.6 \, kg.m^{-3} \quad \mu = 0.008.10^{-3} \, Pa.s$$

$$F = 0.2 \qquad d_p = 60.10^{-6} \, m$$

$$U_{bc} = \frac{2.07 \times 60.10^{-6} \times 2.6^{0.06}}{\left(8.10^{-6}\right)^{0.347}} \exp(0.716 \times 0.2)$$

$$U_{bc} = 0.0089 \, m.s^{-1}$$

We notice that this solid belongs to group A as:

$$U_{bc} \gg U_{fc}$$

EXAMPLE 6.5.–

$$\rho_G = 1.2 \, kg.m^{-3} \qquad \mu = 0.020.10^{-3} \, Pa.s$$

$$F = 0.02 \qquad d_p = 10^{-3} \, m$$

$$U_{bc} = \frac{2.07 \times 10^{-3} \times 1.2^{0.06}}{\left(20.10^{-6}\right)^{0.347}} \exp(0.716 \times 0.02)$$

$$U_{bc} = 0.09 \, m.s^{-1}$$

We notice that for this plant product, the ratio U_{bc}/U_{fc}, equal to 1.4, is much less than that of the catalyst (2.7). Bubbling will therefore be more premature and this solid is very close to group B.

6.3. Morphology of a fluidized bed

6.3.1. *Expansion of a homogenous fluidized bed (without bubbles)*

Let:

$$Z(\varepsilon) = \frac{\varepsilon^3}{1-\varepsilon}$$

ε: porosity of the fluidized bed

We shall write Abrahamsen and Geldart's relation [ABR 80] in the following form:

$$Z(\varepsilon) = Z(\varepsilon_{fc}) + \frac{210(U - U_{fc})\mu_G}{(\rho_S - \rho_G)g\,d_p^2}$$

The porosity is the square root (ranging from 0 to 1) of the third-degree equation.

$$\varepsilon^3 + \varepsilon Z - Z = 0$$

The solution to this type of equation is given in Appendix 4.

The height of the fluidized bed is therefore:

$$H = \frac{M}{A\rho_S(1-\varepsilon)}$$

M: mass of solids: kg

A: horizontal cross-sectional area of the fluidized bed: m²

ρ_S: actual mass density of the particles: kg.m^{-3}

EXAMPLE 6.6.–

$$U = 0.008 \text{ m.s}^{-1} \qquad \varepsilon_{fc} = 0.5 \qquad \rho_S = 1\,100 \text{ kg.m}^{-3}$$

$$U_{fc} = 0.00337 \text{ m.s}^{-1} \qquad d_p = 60.10^{-6} \text{ m} \qquad \rho_G = 2.6 \text{ kg.m}^{-3}$$

$$\mu_G = 0.008.10^{-3} \text{ Pa.s}$$

$$Z(\varepsilon_{fc}) = \frac{0.5^3}{1 - 0.5} = 0.25$$

$$\frac{210(0.008 - 0.00337)0.008.10^{-3}}{1100 \times 9.81 \times (60.10^{-6})^2} = 0.20$$

$$Z(\varepsilon) = 0.25 + 0.20 = 0.45$$

$$p = Z \qquad q = -Z$$

$$\frac{q^2}{4} + \frac{p^3}{27} = \frac{0.45^2}{4} + \frac{(0.45)^3}{27} = 0.054$$

$$\varepsilon = \left[\frac{0.45}{2} + 0.054^{1/2}\right]^{1/3} + \left[\frac{0.45}{2} - 0.054^{1/2}\right]^{1/3} = 0.77045 - 0.194688$$

$$\varepsilon = 0.5757$$

6.3.2. Expansion of the bubbling bed

The porosity relative to the bubbles is:

$$\varepsilon_B = \frac{U - U_{fc}}{u_A}$$

u_A: rate of climb of the bubbles in the fluidized bed: m.s^{-1}

If H_{fc} is the height at incipient fluidization:

$$H_{fc} = (1 - \varepsilon_B)H$$

Let us eliminate ε_B from these two equations:

$$\frac{u_A}{U - U_{fc}} = \frac{H}{H - H_{fc}} \qquad [6.4]$$

Hence the value of H in terms of the flux density in a volume of gas U.

But we know that:

$$u_A = U - U_{fc} + v_B \qquad [6.5]$$

v_B: rate of climb of the bubbles in a stationary gas (m.s^{-1})

Let us eliminate v_A from equations [6.4] and [6.5]. We have

$$\frac{H}{H_{fc}} = 1 + \frac{(U - U_{fc})}{v_B}$$

6.3.3. *Morphology of a bubble*

Traditionally, we assume that a bubble moves with a certain mass of solids which occupies the volume of a spherical cap on the bottom part of the bubble which by analogy with a hot-air balloon, we will call the nacelle although, generally, we prefer to call it a wake.

The sphere describing the bubble therefore has two parts:

– the solid entrained in the nacelle;

– above the nacelle, we have a volume without solid, which is the actual volume of the gas bubble.

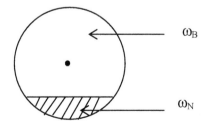

Figure 6.4. *Conventional shape of a bubble*

ω_B: volume of the bubble (m^3)

ω_N: volume of the nacelle (m^3)

$$\omega_N = \frac{\pi h^2}{3}\left(\frac{3}{2}d_{SP} - h\right)$$

h: height of the spherical cap: m

d_{sp}: diameter of the complete sphere

The volume of the actual bubble is:

$$\omega_B = \omega_{SP} - \omega_N = \frac{\pi d_{SP}^2}{6} - \frac{\pi h^2}{3}\left(\frac{3}{2}d_{SP} - h\right)$$

The equivalent diameter of the bubble, that is that of a sphere of same volume with ω_B is:

$$d_{eq} = \left(\frac{6\omega_B}{\pi}\right)^{1/3}$$

The ratio ω_N/ω_{SP}, almost nil for group D particles, decreases from 0.4 to 0.25 when we leave from group A to group B.

Other than the nacelle, the bubble possesses a train whose volume is 15–25% of that of the bubble and which follows the nacelle.

6.3.4. Diameter of bubbles

As we ascend in the fluidized bed, coalescence has a consequence: the diameter of the bubbles increases regularly with the height. Thus, as we draw closer to the surface, the number of bubbles present in a horizontal plane reduces progressively.

Darton *et al.* [DAR 77] gave an expression for the diameter of the bubbles in terms of their height in the fluidized bed. More precisely, it is the equivalent diameter, which is the diameter of the sphere of equal volume as the gaseous part of a bubble.

$$d_{eq} = 0.54 \left(U - U_{fc} \right)^{0.4} \left(h + 4\sqrt{A_o} \right)^{0.8} g^{-0.2} \qquad \text{(International System)}$$

h: height above the distributor (m)

g: acceleration due to gravity (9.81 m.s^{-2})

A_o is the ratio of the surface area of the distributor to the number of supply ports of the gas. We will admit that this definition remains valid whether the gaseous jets are vertical or horizontal. In general, A_o is of the order of 10^{-2} m^2.

EXAMPLE 6.7.–

$$U_{fc} = 0.061 \text{ m.s}^{-1} \qquad h = 0.4 \text{ m} \qquad U = 0.15 \text{ m.s}^{-1}$$

$$U_{bc} = 0.09 \text{ m.s}^{-1} \qquad\qquad A_o = 10^{-2} \text{ m}^2$$

$$d_{eq} = 0.54 \left(0.15 - 0.061 \right)^{0.4} \left(0.4 + 4.10^{-1} \right)^{0.8} \times 9.81^{-0.2}$$

$$d_{eq} = 0.11 \text{ m}$$

NOTE.–

Some authors [HAR 61] had come up with expressions for the maximum diameter above, at which a bubble is unstable, but the diameters obtained are less than those of Darton *et al.* [DAR 77], which correspond to bubbles which really exist and of which we have just given an example.

6.3.5. *Rate of climb of an isolated bubble*

The calculation of this speed involves the equivalent diameter of the bubble:

$$d_{eq} = \left(\frac{6\Omega_B}{\pi} \right)^{1/3}$$

Ω_B: volume of the bubble (m^3)

The Reynolds number is therefore:

$$Re_B = \frac{\rho_F U_B d_{eq}}{\mu_F}$$

ρ_F: mass density of the fluid $(kg.m^{-3})$

μ_F: viscosity of the fluid (Pa.s)

u_B: velocity of the bubble $(m.s^{-1})$

$$u_B = k\sqrt{gd_{eq}}$$

g: acceleration due to gravity $(9.81 \ m.s^{-2})$

Geldart [GEL 86] provides a curve for the variations of k in terms of Re_B:

$$Re_B = \frac{\rho_F u_B d_{eq}}{\mu_F}$$

We shall replace this curve by the following relations. Obviously, the velocity u_B is calculated by iterations.

If $Re_B \leq 10$ k = 0.50

If $Re_B \geq 100$ k = 0.70

If $10 < Re_B < 100$, the value of the coefficient k can be obtained with low error by linear interpolation.

6.3.6. Velocity of the bubbles at a given point of a fluidized bed

For a start, recall that the gas flowrate U_B corresponding to the bubbles is given by:

$$U_B = U - U_{fc}$$

These "flow rates" are taken along the horizontal cross-sectional area of the fluidized bed. They are therefore velocities in an empty bed.

U: overall velocity in an empty bed of the gas (m.s^{-1})

U_{fc}: velocity in an empty bed at the incipient fluidization (m.s^{-1})

In what follows, we shall call the "continuous or dense phase" the portion of the fluidized bed which is exterior to the bubbles (see [NIC 62]).

The cloud rises to the velocity v_n and "swallows" the same amount of continuous phase, which, in the cloud, will have with respect to the bubbles the relative velocity:

$$v_R = \frac{v_n}{1 - \varepsilon_B} \qquad [6.6]$$

This relative velocity is a property of porosity $(1 - \varepsilon_B)$ devolved in the cloud to the continuous phase.

If we now consider the stationary fluidized bed of the same porosity ε_B devolved to the bubbles, the relative velocity of the bubbles with respect to the continuous phase will be:

$$v_R = \frac{U_B}{\varepsilon_B} = \frac{U - U_{fc}}{\varepsilon_B} \qquad [6.7]$$

The bubble cloud and the fluidized bed having the same porosity, the relative velocities are equal and, by eliminating ε_B from the equations [6.6] and [6.7], we get:

$$v_R = v_n + U_B = U - U_{fc} + v_n$$

The velocity v_n is not the rate of climb $v_{B\infty}$ of an isolated bubble but that of a bubble cloud. According to Richardson and Zaki [RIC 54] (see 15.1.5. with $\mu_F = 1$ Pa.s) and according to Latham et al. [LAT 68] the following inequality is verified:

$$v_n < v_{B\infty}$$

6.3.7. Movement of the solid in the fluidized bed

This circulation only really exists if bubbles are present. In fact, they entrain the solid in two different ways:

– in the nacelle;

– in the "train" which follows each bubble.

The solid thus entrained falls back between among the bubbles and both flow rates (upward and downward) are equal. Talmor and Benenati [TAL 63] suggest an expression for the mass flux density corresponding to these rates:

$$W = 785\left(U - U_{fc}\right)e^{-6630d_p}$$

W: flux density of the solid: $kg.m^{-2}.s^{-1}$

d_p: diameter of the particles: m

NOTE.–

Segregation primarily occurs if particles of different mass densities are present. The denser particles come together at the bottom portion of the bed. Segregation can also occur but is less likely if the particles differ by their size. The larger particles accumulate at the bottom portion of the fluidized bed.

Let us note, however, that a voluminous and relatively dense object can "float" on the surface of a fluidized bed, which is explained by the fact that, under the object, particles rise and collide with it while they are absent at the upper portion.

6.3.8. Liberation height of the gas

During the explosion of the bubbles on the surface of the fluidized bed some particles are projected upwards and then fall back on the bed. The maximum height attained by the particles is the liberation height of the gas (LHG.) above which the gas is almost completely free of the solid.

The LHG increases with the gaseous flowrate and also increases with the diameter of the fluidized bed. We will find the corresponding network of curves from Zenz's publication [ZEN 83]. Zenz calls the LHG the TDH in Figure 6. We recall that:

1 in = 2.54 cm   1 ft = 0.3048 m

Zenz's network of curves is usually valid for solid particles of type A and B.

Let us, however, note that Geldart *et al.* [GEL 79] suggested a method for calculating the solid entrained above the LHG by using the distribution of heights of the particles of the fluidized bed as well as their limiting falling velocity.

Finally, Wen and Chen [WEN 82] review the different possible correlations for elutriation (entrainment of solid by the gas).

6.3.9. Calculation of downcomers for superimposed fluidized beds

Adsorption on activated charcoal as well as the regeneration of the latter can be carried out on superimposed fluidized beds in a column. The actual (intrinsic) mass density of the charcoal is low compared to that of the minerals, which limits the difference in gas pressure between the head and foot of the column. This setup allows the countercurrent transfer of matter.

Eleftheriades and Judd [ELE 78] suggested a method for calculating the diameter of a circular weir between two fluidized plates. For this purpose, they use the results from two prior measurements concerning:

– the variation in the fraction of volume of the bed without solid, that is the variation of the porosity of the bed in terms of the gas flow rate;

– the variation in the height of the moving bed in the weir in terms of the gas flow rate in the weir when the beds are just fluidized.

These measures require the installation of a pilot consisting of two fluidized plates as stated by the authors.

6.3.10. The turbulent bed

If the gas flowrate increases in a bubbling bed, the bubbles tear apart and disappear whereas the heterogeneities, after having crossed a maximum, also disappear. These heterogeneities consist of particle conglomerates and veins (or gaseous "tongues"). The concentration of the particles above the surface of the turbulent bed is higher than in the case of the bubbling bed but this surface remains distinct.

Let U_{BT} be the velocity in an empty bed of the gas for the complete transition from the bubbling regime to the turbulent regime. We can draw the following table according to the diameter of the particles d_p.

The geometry of the bed is not considered if the height and diameter of the bed are greater than 2 m. Naturally, the fluidization is more regular if the pressure increases.

15 µm $< d_p <$ 100 µm	0.1 m.s^{-1} $< U_{BT} <$ 0.5 m.s^{-1}
100 µm $< d_p <$ 600 µm	0.5 m.s^{-1} $< U_{BT} <$ 2 m.s^{-1}
600 µm $< d_p$	$U_{BT} >$ 2 m.s^{-1}

Table 6.3. *Bubbling–turbulent transition*

The turbulent regime is adopted in the production of acrylonitrile and the roasting of some sulfides.

The transition of the swab to the rapid (turbulent) regime was defined by Satija and Fan [SAT 85] and a condition for fluidization to take place was established by Yang [YAN 76].

NOTE.–

The simulation of the spouted beds was studied by Smith *et al.* [SMI 82].

6.3.11. *The circulating fluidized bed (transported or rapid)*

In a circulating system the velocity in an empty bed of the gas can reach up to 20 times the limiting falling velocity of an isolated particle.

Under these conditions, the bed would rapidly empty itself of its solid content if the latter was not recycled initially after having crossed an outer cyclone.

The transition between the turbulent regime and rapid regime is made for a velocity in an empty bed of the gas of the order of $2-7$ m.s^{-1} depending on the diameter of the particles. This velocity is what we call the transport velocity.

The strong retromixing explains the fact that the temperature is uniform along the entire height of the rapid bed which, at these velocities, no longer has distinct surface areas. The mass density of the bed is of the order of tens of kg.m^{-3}. Its maximum is at the base and reduces moving higher up through the bed.

6.4. Plugging

6.4.1. Conditions for plugging

Under certain conditions, a bubbling fluidized bed can give rise to the formation of gas pockets, which rise and are separated by dense phase plugs. The dense phase corresponds to the fluidized solid with porosity ε_{fc} for the incipient fluidization.

The pockets as well as the plugs occupy the entire cross-section of the bed.

The plugging can only be fully established if the following relationship exists between the height and diameter of the bed [BAY 74]:

$$H_L \geq 1.34 D^{0.175} = H_C$$

H_L: height of the bed (m)

D: diameter of the bed (m)

But that is not enough. In addition, for the plugging to begin, the velocity of the gas in an empty bed must be equal to:

$$U_{sc} = U_{fc} + 0.07 (g D)^{1/2}$$

Practically, the pockets appear when the diameter of the bubbles is equal to half the diameter of the fluidized bed.

The frequency of the pockets regularly drops with respect to H_{fc} until H_{fc} reaches approximately 1 m. Then, the frequency reaches a limit given by [BAY 74]:

$$f_\ell = \frac{0.35g^{1/2}}{k\,D^{1/2}}$$

f_ℓ: limiting frequency (s^{-1})

D: diameter of the bed (m)

H_{fc}: height of the bed during incipient fluidization (m)

$$k = 1.81\,D^{-0.357}$$

g: acceleration due to gravity (9.81 m.s^{-2})

Bayens and Geldart [BAY 74] define three zones in the fluidized bed each characterized by a height:

– free bubbling zone:

$$h < H_{bl} = \frac{D - 0.063\,D^{0.2}}{1.132\,D^{0.45}}$$

– coalescence zone between bubbles and pockets:

$$H_{bl} < h < H_c = 1.34\,D^{0.175}$$

– established steady swabbing zone:

$$h > H_c$$

EXAMPLE 6.8.–

$D = 1$ m $U_{fc} = 0.061$ m.s^{-1}

$H_c = 1.34$ m

$U_{sc} = 0.061 + 0.07 \times 9.81^{1/2} = 0.28$ m.s^{-1}

$$k = 1.81$$

$$f_1 = \frac{0.35 \times 9.81^{1/2}}{1.81} = 0.6 \text{ s}^{-1} \text{ (or 0.6 Hz)}$$

$$H_{bl} = \frac{1 - 0.063}{1.132} = 0.828 \text{ m}$$

Thus:

free bubbling   for $h < 0.828$ m

coalesced pockets   for $0.828 \text{ m} < h < 1.34$ m

steady pockets   for $h > 1.34$ m

6.4.2. Height of the bed in a plugging (pulsating) regime

When a dense phase plug reaches the bed surface, it progressively disintegrates while it is pushed by the pocket beneath, such that the height of the plug decreases but its upper side continues to rise.

When the plug is close to disappearing, that is when its height draws close to 0, the height of the bed is at its maximum. Then the pocket, which was below the plug, has reached the surface, and the height of the bed suddenly drops from a height equal to that of a pocket.

By analogy with what happens in a bubbling bed (see section 6.2.3), experience shows that [MAT 69]:

$$\frac{H_{max}}{H_{fc}} = 1 + \frac{(U - U_{fc})}{v_s}$$

v_s: characteristic speed of a pocket (m.s^{-1})

$$v_s = 0.35 \text{ (g D)}^{0.5}$$

However, there exist two kinds of pockets:

– those for which the upper side is round and which predominate with solid particles of diameter less than 100 µm. The axis of these pockets merges with that of the bed;

– those for which the upper side is flat and which can exist for particles of size greater than 500 µm. Moreover, some pockets called parietal pockets, for which the axis is moved toward the wall and which no longer occupy the entire cross-section of the bed, can exist for these particles.

The characteristic speed of parietal pockets is therefore:

$$v_{sp} = 0.35\sqrt{2\,g\,D}$$

as indicated by Kehoe and Davidson [KEH 71].

6.4.3. *Fall in pressure across a bed in the swabbing regime [BAY 78]*

This fall in pressure is the sum of two terms:

– one due to the weight of the solid particles $\Delta P_1 = Mg/A$,

– the other due to the acceleration of dense phase plugs:

$$\Delta P_2 = \rho_L f_S \left(H_{fc} - H_S \right) \left(U - U_{fc} + v_S \right)$$

with:

$$H_s = \frac{3.65.10^{-3}\,D^{1.235}}{\left(U - U_{fc} \right)^{1.37}}$$

EXAMPLE 6.9.–

$U_{fc} = 0.061$ m.s^{-1} $H_{fc} = 0.80$ m $U = 0.2$ m.s^{-1}

$f_S = 0.6$ s^{-1}

$\dfrac{Mg}{A} = 4,000$ Pa $\rho_L = 400$ kg.m^{-3} $D = 1$ m

$$H_S = \frac{3.65.10^{-3}}{(0.2-0.061)^{1.37}} = 0.055 \text{ m}$$

$$v_S = 0.35\sqrt{9.81\times1} = 1.097 \text{ m.s}^{-1}$$

$$\Delta P_2 = 400\times0.6\times(0.8-0.055)(0.2-0.061+1.097)$$

$$\Delta P_2 = 236 \text{ Pa}$$

$$\Delta P = 4,000 + 236 = 4,236 \text{ Pa}$$

6.5. Heat transfer

6.5.1. Thermal properties of some solid materials

Table 6.4. gives the properties of materials likely to constitute the particles of a fluidized bed.

	$\lambda(\text{W.m}^{-1}.^{\circ}\text{C}^{-1})$	$C(\text{J.kg}^{-1}.^{\circ}\text{C}^{-1})$	$\rho_s(\text{kg.m}^{-3})$
Alumina	4.5	900	4000
Aluminosilicate	0.36	1060	1200
Limestone	0.95	800	2500
Coal	0.30	1200	1200
Polymers	0.22	1200	550
Sand	1.9	800	2700

Table 6.4. *Thermal properties of materials*

6.5.2. Equivalent thermal conductivity of the fluidized bed

This conductivity is the sum of two terms:

– one essentially due to the agitation of particles by the gas: λ_p [KRU 67]:

$$\ln\left(\frac{\lambda_p}{\lambda_g}\right) = \left[0.28 - 0.757 \log_{10} \varepsilon - 0.057 \log_{10}\left(\frac{\lambda_s}{\lambda_g}\right)\right] \ln\left(\frac{\lambda_s}{\lambda_g}\right)$$

Deissler and Boegli [DEI 58] indicated the effect of the gas pressure.

λ_s: conductivity of the material of which the particles are made $(W.m^{-1}.K^{-1})$

λ_g: conductivity of the gas $(W.m^{-1}.K^{-1})$

ε: porosity of the fluidized bed

– the other due to turbulence caused in the gas by the particles: λ_t

According to Ranz [RAN 52, this author's equation [8]]:

$$\frac{\lambda_t}{\lambda_g} = 0.1 \, Re_p \, Pr$$

with:

$$Re_p = \frac{U d_p \rho_g}{\mu_g} \qquad Pr = \frac{C_g \mu_g}{\lambda_g}$$

U: velocity of the gas in an empty bed: $m.s^{-1}$

d_p: diameter of the particles: m

ρ_g: mass density of the gas: $kg.m^{-3}$

λ_g: conductivity of the gas: $W.m^{-1}K^{-1}$

μ_g: viscosity of the gas: Pa.s

C_g: heat capacity of the gas: $J.kg^{-1}.°C^{-1}$

Finally, the equivalent conductivity of the bed is:

$$\lambda_e = \lambda_t + \lambda_p$$

EXAMPLE 6.10.–

$\varepsilon = 0.5 \quad \lambda_s = 1 \; W.m^{-1}.k^{-1} \quad \lambda_g = 0.035 \; W.m^{-1}.K^{-1}$

$$0.28 - 0.757\log_{10} 0.5 - 0.057\log_{10}\frac{1}{0.035} = 0.4259$$

$$\frac{\lambda_p}{0.035} = \left(\frac{1}{0.035}\right)^{0.4259} \qquad \lambda_p = 0.146 \text{ W.m}^{-1}.\text{K}^{-1}$$

We shall distinguish two cases:

Type A particles:	Types B and C particles:
$d_p = 60.10^{-6}$ m	$d_p = 10^{-3}$ m
$U_{fc} = 0.0036$ m.s^{-1}	$U_{fc} = 0.061$ m.s^{-1}
$\mu_g = 0.008.10^{-3}$ Pa.s	$\mu_g = 0.020.10^{-3}$ Pa.s
$\lambda_g = 0.035$ W.m^{-1}.K^{-1}	$\lambda_g = 0.035$ W.m^{-1}.K^{-1}
$\rho_g = 2.6$ kg.m^{-3}	$\rho_g = 1.2$ kg.m^{-3}
Pr = 0.7	Pr = 0.7

$$\text{Re}_p = \frac{0.0036 \times 60.10^{-6} \times 2.6}{0.008.10^{-3}} \qquad \text{Re}_p = \frac{0.061 \times 10^{-3} \times 1.2}{20.10^{-6}}$$

$\text{Re}_p = 0.0702$	$\text{Re}_p = 3.66$
$\lambda_t = 0.1 \times 0.035 \times 0.0702 \times 0.7$	$\lambda_t = 0.1 \times 0.035 \times 3.66 \times 0.7$
$\lambda_t = 1.72.10^{-4}$	$\lambda_t = 89.67.10^{-4}$
$\lambda_e = 0.146 + 1.72.10^{-4}$	$\lambda_e = 0.146 + 89.67.10^{-4}$
$\lambda_e = 0.146172$ W.m^{-1}.K^{-1}	$\lambda_e = 0.155$ W.m^{-1}.K^{-1}

6.5.3. Heat transfer (by convection) across a vertical surface

The coefficient for the convection is the sum of two terms:

1) a term, which is relative to the particles:

Type A particles ([GAB 70a, GAB 70b], this author's equation [11]);

$$\alpha_p = 2\left[\frac{\lambda_e \rho_e C_e \left(U - U_{fc}\right)}{\pi L}\right]^{1/2}$$

Types B and D particles [BOT 81];

$$\alpha_p = 0.84\frac{\lambda_g}{d_p} Ar^{0.15} \text{ with } Ar = \frac{\rho_g \left(\rho_s - \rho_g\right)g\, d_p^3}{\mu_g^2}$$

ρ_e: equivalent mass density of the homogenous portion of the bed (kg.m^{-3})

$$\rho_e = \left(1 - \varepsilon_{fc}\right)\rho_s + \varepsilon_{fc}\rho_g$$

C_e: equivalent heat capacity of the homogenous portion of the bed (J.kg^{-1}.°C^{-1})

$$C_e = \frac{C_s \rho_s \left(1 - \varepsilon_{fc}\right) + C_g \rho_g \varepsilon_{fc}}{\rho_e}$$

L is the smallest of the following measurements (in meters):

– diameter of bubbles,

– vertical dimension of the surface.

d_p: diameter of the particles (m)

μ_g: viscosity of the gas (Pa.s)

But the particles are not in direct contact with the surface and it is necessary to introduce resistance corresponding to a gas film.

$$\frac{1}{\alpha_{fi}} = \frac{d_p}{6\lambda_g}$$

Hence:

$$\alpha_{pc} = \frac{(1-\varepsilon_B)}{\dfrac{1}{\alpha_p} + \dfrac{1}{\alpha_{fi}}}$$

ε_B: volume fraction of the fluidized bed occupied by the bubbles

2) a term which is relative to the gas:

According to Botterill and Denloye [BOT 78]:

$$\alpha_{gc} = 0.863 \frac{\lambda_g}{d_p^{1/2}} Ar^{0.39}$$

with:

$$10^3 < Ar < 2.10^6$$

3) the overall coefficient for the convection is therefore:

$$\alpha_c = (1-\varepsilon_B)(\alpha_{pc} + \alpha_{gc})$$

EXAMPLE 6.11.–

For both types of solid:

$\varepsilon_B = 0.2$ $\rho_s = 1200 \ kg.m^{-3}$ $\varepsilon_{fc} = 0.5$

$\lambda_g = 0.035 \ W.m^{-1}.K^{-1}$

Type A solid   *Types B and D solid*

$\lambda_e = 0.146172 \ W.m^{-1}.K^{-1}$ $\lambda_e = 0.155 \ W.m^{-1}.K^{-1}$

$U_{fc} = 0.0036 \ m.s^-$ $U_{fc} = 0.061 \ m.s^{-1}$

$d_p = 60.10^{-6} \ m$ $d_p = 10^{-3} \ m$

$\rho_s = 1200 \ kg.m^{-3}$

$\rho_g = 2.6 \ kg.m^{-3}$ $\rho_g = 1.2 \ kg.m^{-3}$

$U_{bc} = 0.0089 \text{ m.s}^{-1}$ $U_{bc} = 0.09 \text{ m.s}^{-1}$

$\rho_e = 600 \text{ kg.m}^{-3}$ $\mu_g = 0.020.10^{-3} \text{ Pa.s}$

$C_e = 1000 \text{ J. kg.}°C^{-1}$ $\rho_s = 1,200 \text{ kg.m}^{-3}$

$L = 0.2 \text{ m}$ $Ar = \dfrac{1.2(1,200-1.2)\times 9.81\times 10^{-9}}{\left(0.020.10^{-3}\right)^2}$

$\mu_g = 0.008.10^{-3} \text{ Pa.s}$ $Ar = 35,316$

$U = 0.011 \text{ m.s}^{-1}$ $\alpha_p = \dfrac{0.84\times 0.035}{10^{-3}}\times 35,316^{0.15}$

$$\alpha_p = 2\left[\frac{0.146\times 600\times 1,000(0.011-0.0036)}{\pi\times 0.2}\right]^{1/2} \qquad \alpha_p \quad = \quad 141.4$$
$W.m^{-2}.°C^{-1}$

$\alpha_p = 64.23 \text{ W.m}^{-2}.°C^{-1}$

$\alpha_{pc} = \dfrac{(1-0.2)}{\dfrac{1}{64.23}+\dfrac{60.10^{-6}}{6\times 0.035}} = 50.45$ $\alpha_{pc} = \dfrac{(1-0.2)}{\dfrac{1}{141.4}+\dfrac{10^{-3}}{6\times 0.035}} = 67.60$

$Ar = \dfrac{2.6(1200-2.6)\times 9.81\times \left(60.10^{-6}\right)^3}{\left(0.008.10^{-3}\right)^2}$ $\alpha_{gc} = \dfrac{0.863\times 0.035}{\left(10^{-3}\right)^{1/2}}\left(35,316\right)^{0.39}$

$Ar = 103.30$ $\alpha_{gc} = 56 \text{ W.m}^{-2}.°C^{-1}$

$\alpha_{gc} = \dfrac{0.863\times 0.035\times 103.30^{0.39}}{\left(60.10^{-6}\right)^{1/2}}$

$\alpha_{gc} = 23.3 \text{ W.m}^{-2}.°C^{-1}$

$\alpha_c = (1-0.2)(46.76+23.3)$ $\alpha_c = (1-0.2)(141.4+56)$

$\alpha_c = 56.05 \text{ W.m}^{-2}.°C^{-1}$ $\alpha_c = 157.92 \text{ W.m}^{-2}.°C^{-1}$

6.5.4. *Heat transfer between a fluidized bed and a vertical tube*

Genetti and Knudsen [GEN 68] suggest the following expression for the transfer coefficient (or, rather, for the Nusselt number):

$$Nu = \frac{5\phi(1-\varepsilon)^{0.48}}{\left[1+\dfrac{580}{60\,Re}\left(\dfrac{\lambda_s}{d_p^{3/2}C_s\rho_s g^{1/2}}\right)\left(\dfrac{\rho_s}{\rho_g}\right)^{1.1}\left(\dfrac{U_{fc}}{U}\right)^{4/3}\right]^2}$$

Re: Reynolds number of particles

$$Re = \frac{U d_p \rho_g}{\mu_g} \qquad Nu = \frac{d_p \alpha}{\lambda_g}$$

μ_g: viscosity of the gas: Pa.s

The coefficient 60 comes from the fact that the authors use the hour as the unit of time everywhere, except for the acceleration due to gravity g where they use the second. Thus, the formula is homogenous (and dimensionless).

ρ_g and ρ_s: mass densities of the gas and the solid: $kg.m^{-3}$

λ_s: conductivity of the solid ($W.m^{-1}.°C^{-1}$)

C_s: heat capacity of the solid ($J.kg^{-1}.°C^{-1}$)

g: acceleration due to gravity ($m.s^{-2}$)

U_{fc}: velocity of the gas in an empty bed at incipient fluidization ($m.s^{-1}$)

ϕ: coefficient of nonsphericity of the particles (see section 6.1.5)

$$\phi = \frac{\text{surface area of a particle}}{\text{surface area of the sphere of same volume as the particle}}$$

ε: porosity of the fluidized bed

If the tube is inserted in a bundle of vertical tubes, the transfer coefficient α will vary by less than 20%.

EXAMPLE 6.12.–

$\rho_s = 1100 \text{ kg.m}^{-3}$ $\mu_g = 0.020.10^{-3} \text{ Pa.s}$ $\epsilon = 0.5$

$\rho_g = 2.6 \text{ kg.m}^{-3}$ $g = 9.81 \text{ m.s}^{-2}$ $U_{fc}/U = 1$

$d_p = 60.10^{-6} \text{ m}$ $U_{fc} = 0.0089 \text{ m.s}^{-1}$ $\phi = 1$

$c_s = 1000 \text{ J.kg.}^\circ\text{C}^{-1}$

$$Re = \frac{0.0089 \times 60.10^{-6} \times 1{,}100}{0.020.10^{-3}} = 29.37$$

$$1/\left[\left(60.10^{-6}\right)^{1.5} \times 1{,}000 \times 1{,}100 \times 9.81^{0.5}\right] = 0.6246$$

$$\left[1 + \frac{580}{60 \times 29.37} \times 0.035 \times 0.6246\left(\frac{1{,}100}{2.6}\right)^{1.1}\right]^2 = 43.20$$

$$Nu = \frac{5 \times 0.5^{0.48}}{43.20} = 0.0829$$

$$\alpha = \frac{0.035 \times 0.0829}{60.10^{-6}} = 48.35 \text{ W.m}^{-2}.^\circ\text{C}^{-1}$$

6.5.5. *Heat transfer between the fluidized bed and a horizontal tube*

The tube's Nusselt number is given by [VRE 58]:

$$Nu = 420\left(\frac{G\,D_T\rho_s}{\rho_G\mu_G} \times \frac{\mu_G^2}{d_p^3\rho_s^2 g}\right)^{0.3} Pr^{0.3}$$

with:

$$Nu = \frac{\alpha D_T}{\lambda_G} \qquad Pr = \frac{C_G \mu_G}{\lambda_G} \qquad G = U\rho_G$$

α: heat transfer coefficient: $W.m^{-2}.°C^{-1}$

D_T: outer diameter of the tube

ρ_S and ρ_G: mass densities of the solid and the gas: $kg.m^{-3}$

μ_G: viscosity of the gas: Pa.s

U: velocity of the gas in an empty bed. $m.s^{-1}$

C_G: heat capacity of the gas: $J.kg.°C$

λ_G: thermal conductivity of the gas: $W.m^{-1}.°C^{-1}$

Pr: Prandtl number

The coefficient of an isolated horizontal tube transported in a bundle of horizontal tubes does not vary by more than 20%.

EXAMPLE 6.13.–

$U = 0.008 \text{ m.s}^{-1}$ $\mu_g = 0.008.10^{-3}.\text{Pa.s}$ $\rho_s = 1100 \text{ kg.m}^{-3}$

$D_T = 0.025 \text{ m}$ $g = 9.81 \text{ m.s}^{-2}$ $\rho_G = 2.6 \text{ kg.m}^{-3}$

$d_p = 60.10^{-6} \text{ m}$ $\lambda_G = 0.036 \text{ W.m}^{-1}.°C^{-1}$ $Pr = 0.7$

$$Nu = 420 \left[\frac{0.008 \times 2.6 \times 0.025 \times 1,100}{2.6 \times 0.008.10^{-3}} \times \frac{\left(0.008.10^{-3}\right)^2}{\left(60.10^{-6}\right)^3 \times 1,100^2 \times 9.81} \times 0.7 \right]^{1/3}$$

$Nu = 420 \times 1.60 = 672$

$$\alpha = \frac{672 \times 0.036}{0.025} = 968 \text{ W.m}^{-2}.°C^{-1}$$

6.5.6. *Heat transfer by radiation*

The heat load transferred is:

$$\phi_r = \sigma\varepsilon\left(T_C^4 - T_F^4\right) \qquad \left(W.m^{-2}\right)$$

σ: Stefan constant: $5{,}673.10^{-8}$ $W.m^{-2}.K^{-4}$

T_C: absolute temperature of the hot surface: K

T_F: absolute temperature of the cold surface: K

ε: apparent emissivity for the system

$$\varepsilon = \cfrac{1}{\cfrac{1}{\varepsilon_L} + \cfrac{1}{\varepsilon_S} - 1}$$

ε_L: emissivity of the fluidized bed (equal to that of the particles)

ε_S: emissivity of the surface in contact with the bed

Nature of the surface	ε
Polished metal mirror	0.05
Stainless steel	0.2
Mild steel, cast iron	0.3
Sand	0.5
Alumina	0.65

Table 6.5. *Practical values of the emissivity*

NOTE.–

We know that:

$$T_C^4 - T_F^4 = \left(T_C - T_F\right)\left(T_C^3 + T_F^3 + T_C^2 T_F + T_C T_F^2\right)$$

We can then write:

$$\phi_r = \alpha_r \left(T_C - T_F \right)$$

with:

$$\alpha_r = \sigma\varepsilon \left(T_C^3 + T_F^3 + T_C^2 T_F + T_C T_F^2 \right)$$

6.6. Applications of fluidization

All these applications result from the fact that in a fluidized bed the contact between the fluid and the solid particles is very intense (see 11.3.1).

Thus, fluidization is very convenient:

– for the drying of crystals;

– for reactions in the gaseous phase in the presence of a solid catalyst;

– for the coating of particles as the solvent is rapidly evaporated by the gas stream [GUT 91].

Air slides are chutes closed over their entire cross-sectional area (which is square) which are inclined and where the divided solid is kept in the fluidized state by injection of air across the lower side (which is porous). Botterill and Bessant [BOT 73] investigated the behavior of the divided solid in the chute.

APPENDICES

Appendix 1

Apparent Mass Density of Bulk Divided Solids (kg.m^{-3})

A1.1. Vegetable products

Nature of product	Grains		Flours	
Linen	720		430	
Corn	720		640	
Cotton	530		400	
Soy	700	(shredded)	540	
Coffee	670	(green)	450	(roasted, ground)
Wheat	790			
Barley	620			
Rye	720			
Rice	800			
Oats	410			
Clove	770			

A1.2. Natural inorganic products

Nature of product	Grained		Powders (fine grinding)	
Bauxite	1,360	(bulk)	1,090	
Gypsum	1,270		900	
Kaolin	1,024	(crushed)	350	(< 10 μm)
Lead silicate	3,700		2,950	
Quicklime	850		430	
Limestone	1,570		1,360	
Phosphate	960		800	

Wood waste	350	(shavings)	320	(sawdust)
Sulphur	1,220		800	
Iron	4,950	(balls)	2,370	(filings)
Schist slate	1,390		1,310	
Sodium carbonate	1,060		480	
Coke	490		430	

A1.3. Manufactured products

Powdered sodium bicarbonate	690
Borax	1,700
Catalyst (fluidized petroleum cracking)	510
Ashes	700
Charcoal (grains)	420
Coal (bulk)	900
Coal (rated)	700–800
Cement (clinker)	1,400
Cement (Portland)	1,520
Shredded copra shavings	510
Copra shavings with presser screw outlet	465
Dolomite powder	730
Soap flakes	160
Crushed feldspar	1,600
Gravel	1,500
Dairy	2,000
Milled mica	210
Crushed bone	1,200
Phthalic anhydride flakes	670
Glass beads	1,400
Iron oxide pigment	400
Zinc oxide pigment	320
Lead shot	6,560
Potatoes	700
Rubber trimmings	370
Sand	1,350–1,500
Salt	1,200
Granulated sugar	830
Crystallized copper sulphate	1,200
Superphosphate powder	810

Appendix 2

Simple Results of Analytical Geometry

A2.1. Product of two vectors V_1 and V_2

Let us consider a cube corner for which all three axes Ox, Oy, and Oz are oriented like the thumb, forefinger, and the middle finger. Let i_x, i_y, and i_z respectively be the unit vectors along Ox, Oy, and Oz. On the other hand, let x_1, y_1, and z_1 be the coordinates of V_1 and x_2, y_2, and z_2 the coordinates of V_2.

1) Inner product, that is scalar product:

the product is represented by a dot. We shall write:

$$i_x.i_x = 1 \qquad i_y.i_y = 1 \qquad i_z.i_z = 1$$

$$i_x.i_y = 0 = i_y.i_x \qquad i_x.i_z = 0 = i_z.i_x \qquad i_y.i_z = 0 = i_z.i_y$$

$$V_1 = i_x x_1 + i_y y_1 + i_z z_1 \quad V_2 = i_2 x_2 + i_y y_2 + i_z z_2$$

hence:

$$V_1.V_2 = x_1 x_2 + y_1 y_2 + z_1 z_2$$

2) Outer product, that is vector product:

the product is simply represented by the "multiplication" symbol.

We shall write:

$$i_x \times i_y = i_z \qquad {}_y \times i_x = -i_z \qquad i_x \times i_x = 0$$

$$i_y \times i_z = i_x \qquad i_z \times i_y = -i_x \qquad i_y \times i_y = 0$$

$$i_z \times i_x = i_y \qquad i_x \times i_z = -i_y \qquad i_z \times i_z = 0$$

Let us consider the two vectors $\overrightarrow{V_1}$ and $\overrightarrow{V_2}$:

$$\overrightarrow{V_1} = x_1 i_x + y_1 i_y + z_1 i_z = i_x x_1 + i_y y_1 + i_z z_1$$

$$\overrightarrow{V_2} = x_2 i_x + y_2 i_y + z_2 i_z = i_x x_2 + i_y y_2 + i_z z_2$$

The outer product of these two vectors is:

$$
\begin{aligned}
\overrightarrow{V_1} \times \overrightarrow{V_2} &= x_1 \left(i_x \times i_x x_2 + i_x \times i_y y_2 + i_x \times i_z z_2 \right) \\
&+ y_1 \left(i_y \times i_x x_2 + i_y \times i_y y_2 + i_y \times i_z z_2 \right) \\
&+ z_1 \left(i_z \times i_x x_2 + i_z \times i_y y_2 + i_z \times i_z z_2 \right) \\
&= x_1 \left(i_z y_2 - i_y z_2 \right) + y_1 \left(-i_z x_2 + i_x z_2 \right) \\
&+ z_1 \left(i_y x_2 - i_x y_2 \right) = \overrightarrow{i_x} \left(y_1 z_2 - z_1 y_2 \right) \\
&+ \overrightarrow{i_y} \left(z_1 x_2 - x_1 z_2 \right) + \overrightarrow{i_z} \left(x_1 y_2 - y_1 x_2 \right)
\end{aligned}
$$

The coefficients of $\overrightarrow{i_x}, \overrightarrow{i_y}, \overrightarrow{i_z}$ are the coordinates of $\overrightarrow{V_1} \times \overrightarrow{V_2}$ in the cube corner.

A2.2. The rotation vector Ω and the velocity vector

The vector with coordinates ω_x, ω_y, ω_z defines the rotation of a point M away from the axis carrying the vector Ω. This distance OM has r_x, r_y, and r_z as coordinates.

The velocity vector V of the point M is:

$$V = \Omega \times R = i_x \left(\omega_y r_z - \omega_z r_y \right) + i_y \left(\omega_z r_x - \omega_x r_z \right) + i_z \left(\omega_x r_y - \omega_y r_x \right)$$

The vector V is the circumferential velocity of the point M.

Some authors write this outer product as a matrix product:

$$V = \begin{pmatrix} 0 & -\omega_z & \omega_y \\ \omega_z & 0 & -\omega_x \\ -\omega_y & \omega_x & 0 \end{pmatrix} \begin{pmatrix} r_x \\ r_y \\ r_z \end{pmatrix}$$

A2.3. Normal to a plane

The most general equation of the plane is:

$$ax + by + cz = q$$

Let us find the equations defining the straight line perpendicular to E from the point M_0 (x_0, y_0, z_0) on the plane P. This straight line is parallel to the vector V (a, b, c), which, in turn, is perpendicular to the plane P.

$$\frac{x - x_0}{a} = \frac{y - y_0}{b} = \frac{z - z_0}{c} = k$$

Or:

$$x = x_0 + ak \quad y = y_0 + bk \quad z = z_0 + ck \qquad \text{[A2.1]}$$

Let us now substitute these three values in the equation of the plane P in order to determine the coordinates x_z, y_x, and $z_?$ of the point H at the foot of the perpendicular from M_0 on P.

$$ax_0 + by_0 + cz_0 + k \left(a^2 + b^2 + c^2 \right) = q$$

Hence:

$$k = \frac{q - \left(ax_0 + by_0 + cz_0 \right)}{a^2 + b^2 + c^2}$$

The numerator of k is not nil because the point M_o is not contained in the plane P.

The equations [A2.1] provide the coordinates of the point H.

The orthogonal projection of a vector or a segment of a straight line is obtained by projecting each of the extreme points.

Appendix 3

Mohs' Scale

Nature of divided solid	Mohs' index
Wax	0.02
Graphite	0.5–1
Talc	1
Diatomaceous earth	1–1.5
Asphalt	1.5
Lead	1.5
Gypsum	2
Human nail	2
Organic crystals	2
Soda flakes	2
Slaked lime	2–3
Sulphur	2
Salt	2
Tin	2
Zinc	2
Anthracite	2.2
Silver	2.5
Borax	2.5
Kaolin	2.5
Litharge	2.5
Baking soda	2.5
Copper (coins)	2.5

Slaked lime	2–3
Aluminum	2–3
Quicklime	2–4
Calcite	3
Bauxite	3
Mica	3
Plastic materials	3
Barite	3.3
Brass	3–4
Limestone	3–4
Dolomite	3.5–4
Siderite	3.5–4
Sphalerite	3.5–4
Chalcopyrite	3.5–4
Fluorite	4
Pyrrhotite	4
Iron	4–5
Zinc oxide	4.5
Glass	4.5–6.5
Apatite	5
Carbon black	5
Asbestos	5
Steel	5–8.5
Chromite	5.5
Magnetite	6
Orthoclase	6
Clinker	6
Iron oxide	6
Feldspar	6
Pumice	6
Magnesia (MgO)	5–6.5
Pyrite	6.5
Titanium oxide	6.5
Quartz	7

Sand	7
Zirconia	7
Beryl	7
Topaz	8
Emery	7–9
Garnet	8.2
Sapphire	9
Corundum	9
Tungsten carbide	9.2
Alumina	9.25
Tantalum carbide	9.3
Titanium carbide	9.4
Silicon carbide	9.4
Boron carbide	9.5
Diamond	10

We can also classify materials according to their hardness:

Tender   1–3

Somewhat tender   4–6

Hard   7–10

Bibliography

[ABR 80] ABRAHAMSEN A.R., GELDART D., "Behaviour of gas-fluidized beds of fine powders. Part I. Homogeneous expansion", *Powder Technology*, vol. 26, p. 35, 1980.

[ADA 05] ADAMIEC P., NEMOZ-GAILLARD M., LEPETZ D. *et al.*, "Etude du comportement mécanique de milieux granulaires modèles", *Récents Progrès en Génie des Procédés*, vol. K-6, no. 92, pp. 1–8, 2005.

[AND 67a] ANDERSON T.B., JACKSON R., "Hydrodynamic stability of a fluidized bed", *Industrial and Engineering Chemistry Fundamentals*, vol. 6, no. 3, pp. 478–480, 1967.

[AND 67b] ANDERSON T.B., JACKSON R., "A fluid mechanical description of fluidized beds", *Industrial and Engineering Chemistry Fundamentals*, vol. 6, pp. 527–539, 1967.

[AUL 77] AULTON M.E., "Micro-indentation tests for pharmaceuticals", *Manufacturing Chemist and Aerosol News*, p. 28, May, 1977.

[AUL 81] AULTON M.E., MAROK I.S., "Assessment of the work-hardening characteristics of some tabletting materials using Meyer's relationship", *International Journal of Pharmaceutical Technology & Product Manufacture*, vol. 2, no. 1, p. 1, 1981.

[BAG 54] BAGNOLD F.R.S., "Experiments on a gravity-free dispersion of large solid spheres in a Newtonian fluid under shear", *Proceedings of the Royal Society of London, Series A: Mathematical and Physical Sciences*, vol. 225, pp. 49–63, 1954.

[BAR 77] BARQUIN M., MAUGIS D., COURTEL R., "Rôle de l'adhésion dans le frottement du caoutchouc. Nouvelle définition du coefficient de frottement", *Compte-rendu de l'Académie des Sciences de Paris. Série B: Mécanique Physique*, vol. 284, pp. 127–130, 1977.

[BAT 56] BATEL W., Forschung Bericht. Nr 262 Witschafts und Verkehrsmin. Nordrh. Westfal. Düsseldorf, 1956.

[BAT 88a] BATCHELOR G.K., "A new theory of the instability of a uniform fluidized bed", *Journal of Fluid Mechanics*, vol. 193, p. 75, 1988.

[BAT 88b] BATHURST R.J., ROTHENBURG L., "Micromechanical aspects of isotropic granular assemblies with linear contact interactions", *Journal of Applied Mechanics*, vol. 55, pp. 17–23, 1988.

[BAY 73] BAYENS J., GELDART D., "Predictive calculations of flow parameters in gas fluidized beds and fluidization behaviour of various powders", *Colloque international sur la fluidisation et ses applications,* Toulouse, France, pp. 263–273, 1–5 October 1973.

[BAY 74] BAYENS J., GELDART D., "An investigation into slugging fluidized beds", *Chemical Engineering Science*, vol. 29, p. 255, 1974.

[BAY 78] BAYENS J., GELDART D., "An investigation into the slugging characteristics of large particles", *Powder Technology*, vol. 19, p. 177, 1978.

[BEL 94] BELL T.A., ENNIS B.J., GRYGO R.J. et al., "Practice evaluation of the Johanson Hang-up indicizer", *Bulk Solids Handling*, vol. 14, pp. 117–125, 1994.

[BEV 61] BEVERLOO W.A., LENIGER H.A., VAN DE VELDE J., "The flow of granular solids through orifices", *Chemical Engineering Science*, vol. 15, pp. 260–269, 1961.

[BOT 73] BOTTERILL J.S.M., BESSANT D.J., "The flow properties of fluidized solids", *Powder Technology*, vol. 8, pp. 213–222, 1973.

[BOT 78] BOTTERILL S.M., DENLOYE A.O.O., "Gas convective heat transfer to packed and fluidized beds", *AIChE Symposium Series*, vol. 74, p. 194, 1978.

[BOT 81] BOTTERILL J.S.M., YEOMAN Y., YUREGIR K.R., "Temperature effects on the heat transfer behaviour of gas fluidized beds", *AIChE Symposium Series*, vol. 77, no. 208, p. 330, 1981.

[BOU 94] BOUCHAUD J.P., CATES M.E., PRAKASH J.R. et al., "A model for the dynamics of sandpile surfaces", *Journal de Physique*, vol. 4, pp. 1383–1410, 1994.

[BOU 97] BOUTREUX T., DE GENNES P.G., "Etalement d'une marche de sable: le problème du Sinaï", *Compte-rendus de l'Académie des Sciences de Paris. Série IIb, Solides: propriétés mécaniques et thermiques*, vol. 325, pp. 85–89, 1997.

[BOU 98] BOUTREUX T., DE GENNES P.G., "Arrêt d'un écoulement granulaire sur une pente faible", *Compte-rendus de l'Académie des Sciences de Paris Série IIb Solides; propriétés mécaniques et thermiques*, vol. 326, pp. 257–262, 1998.

[BRO 50] BROWN G.G. *et al.*, *Unit Operations*, John Wiley & Sons, New York, 1950.

[BRO 61] BROWN R.L., "Minimum energy theorem for flow of dry granules through apertures", *Nature*, vol. 191, pp. 458–481, 1961.

[BRO 65] BROWN R.L., RICHARDS J.C., "Kinematics of the flow of dry powders and bulk solids", *Rheologica Acta*, vol. 4, no. 3, pp. 153–165, 1965.

[BUL 64] BULSARA P.U., ZENZ F.A., ECKERT R.A., "Pressure and additive effects on flow of bulk solids", *Industrial and Engineering Chemistry Process Design and Development*, vol. 3, no. 4, pp. 348–355, 1964.

[CAM 85] CAMPBELL C.S., BRENNEN C.E., SABERSKY R.H., "Flow regimes in inclined open-channel flows of granular materials", *Powder Technology*, vol. 41, pp. 77–82, 1985.

[CAM 01] CAMBOU B., JEAN M., *Micromécanique des matériaux granulaires*, Hermes Science, Paris, 2001.

[CAR 65a] CARR R.L., "Evaluating flow properties of solids", *Chemical Engineering*, vol. 72, no. 2, pp. 163–168, 1965.

[CAR 65b] CARR R.L., "Classifying flow properties of solids", *Chemical Engineering*, vol. 72, no. 3, pp. 69–72, 1965.

[CAR 68] CARR J.F., WALKER D.M., "An annular shear cell for granular materials", *Powder Technology*, vol. 1, pp. 369–373, 1967–1968.

[CAR 70] CARR R.L., "Particle behavior storage and flow", *British Chemical Engineering*, vol. 15, no. 12, pp. 1541–1549, 1970.

[CAR 94] CARSON J.W., MARINELLI J., "Characterize bulk solids to ensure smooth flow", *Chemical Engineering*, vol. 101, no. 4, pp. 78–90, 1994.

[CHA 89] CHANG C.S., MISRA A., XUE J.H., "Incremental stress–strain relationships for regular packings made of multi-sized particles", *International Journal of Solids Structures*, vol. 25, no. 6, pp. 665–681, 1989.

[COU 73] COULOMB C.A., "Essai sur une application des règles de maximis et minimis à quelques problèmes de statique, relatifs à l'architecture", *Mémoires de mathématiques et de physique présentés à l'Académie Royale des Sciences par divers savants et lus dans ses assemblées*, vol. 7, pp. 343–382, 1773.

[COU 00] COURUYER C., NING Z., GHADIRI M., "Distinct element analysis of bulk crushing: effect of particle properties and loading rate", *Powder Technology*, vol. 109, pp. 241–254, 2000.

[CUN 79] CUNDALL P.A., STRACK O.D.L., "A discrete numerical model for granular assemblies", *Géotechnique*, vol. 29, no. 1, pp. 47–65, 1979.

[DAR 77] DARTON R.C., LA NAUZE R.D., DAVIDSON J.F. *et al.*, "Bubble growth due to coalescence in fluidised beds", *Transactions of the Institution of Chemical Engineers*, vol. 55, p. 274, 1977.

[DAV 73] DAVIDSON J.F., NEDDERMAN R.M., "The hour-glass theory of hopper flow", *Transactions of the Institution of Chemical Engineers*, vol. 51, pp. 29–35, 1973.

[DAV 77] DAVIDSON J.F., HARRISON D., GUEDES DE CARVALHO J.R.F., "On the liquidlike behaviour of fluidized beds", *Annual Review of Fluid Mechanics*, vol. 9, p. 55, 1977.

[DAV 85] DAVIDSON J.F., CLIFT R., HARRISON D., *Fluidization*, 2nd ed., Academic Press, 1985.

[DEG 95] DE GENNES P.G., "Dynamique superficielle d'un matériau granulaire", *Comptes-rendus de l'Académie des Sciences de Paris. Série IIb. Physique des surfaces et des interfaces*, vol. 321, pp. 501–506, 1995.

[DEG 97] DE GENNES P.G., "Avalanches of granular materials", in MALLAMACE F., STANLEY H.E. (eds), *Physics of Complex Systems*, Enrico Mermi-Editions LOS Press, Amsterdam, 1997.

[DEH 85] DEHNE D., KELLER H., "Untersuchung des Einflusses von Schüttgut und Bunkerparametern auf die Gesetzmässigkeiten des Ausflusses kohäsionsloser Schüttgüter aus Schlitzbunkern", *Chemie and Technik*, vol. 37, pp. 420–421, 1985.

[DEI 58] DEISSLER R.G., BOEGLI J.S., "An investigation of effective thermal conductivities of powders in various gases", *Transactions of the American Society of Mechanical Engineers*, vol. 80, p. 1417, 1958.

[DEL 56] DELAPLAINE J.W., "Forces acting in flowing beds of solids", *Journal of American Institute of Chemical Engineers*, vol. 2, no. 1, pp. 127–138, 1956.

[DEV 75] DEVISE B., DELACOURTE-THIBAUT A., GUYOT J.C. *et al.*, "Mise au point d'une technique d'étude simplifiée de l'écoulement des poudres destinées à la compression", *Pharmaceutica Acta*, vol. 50, no. 12, pp. 432–446, 1975.

[DON 84] DONSI G. *et al.*, "On a test arrangement of a glow-discharge probe for fluid bed diagnostics", *Powder Technology*, vol. 39, pp. 15–19, 1984.

[DRE 85] DRESCHER A., VGENOPOULOU I., "A theoretical analysis of channeling in bins and hoppers", *Powder Technology*, vol. 42, pp. 181–191, 1985.

[DRU 51] DRUCKER D.C., "A more fundamental approach to plastic stress–strain relations", *Proceedings of the first U.S. National Congress of Applied Mechanics*, pp. 487–491, 11–16 June 1951.

[DUC 85] DUCKER J.R., DUCKER M.E., NEDDERMAN R.M., "The discharge of granular materials from unvenilated hoppers", *Powder Technology*, vol. 42, pp. 3–14, 1985.

[DUR 99] DUROUDIER J.P., *Pratique de la filtration*, Hermès, 1999.

[DUR 02a] DURU P., GUAZZELLI E., "Experimental investigation on the secondary instability of liquid-fluidized beds and the formation of bubbles", *Journal of Fluid Mechanics*, vol. 470, p. 359, 2002.

[DUR 02b] DURU P., NICOLAS M., HINCH J. *et al.*, "Constitutive laws in liquid-fluidized beds", *Journal of Fluid Mechanics*, vol. 452, p. 371, 2002.

[DUR 16] DUROUDIER J.-P., *Divided Solid Mechanics*, ISTE Press Ltd, London and Elsevier Ltd, Oxford, 2016.

[ELE 78] ELEFTHERIADES C.M., JUDD M.R., "The design of downcomers joining gas-fluidized beds in multistage systems", *Powder Technology*, vol. 21, pp. 217–225, 1978.

[ENS 75] ENSTAD G., "On the theory of arching in mass flow hoppers", *Chemical Engineering Science*, vol. 30, pp. 1273–1283, 1975.

[ENS 77] ENSTAD G., "A note on the stresses and dome formation in axially symmetric mass flow hoppers", *Chemical Engineering Science*, vol. 32, pp. 337–339, 1977.

[ERG 51] ERGUN S., "Determination of particle density of crushed porous solids. Gas flow method", *Analytical Chemistry*, vol. 23, pp. 151–156, 1951.

[ERL 71] ERLE M.A., DYSON D.C., MORROW N.R., "Liquid bridges between cylinders, in a torus and between sphères", *AIChE Journal*, vol. 17, no. 1, p. 115, 1971.

[FER 81] FERRY R., MICHALET R., PROVOST P., *Dictionnaire pratique de physique*, G.E.D.I.C., 1981.

[FIR 84] FIREWICZ H., "Kinetik des Schwerkraft-Auslaufs körnigen Gutes aus einem Bunker", *Aufbereitungs-Technik*, vol. 25, pp. 209–214, 1984.

[FIR 85] FIREWICZ H., "Kinetik des Schwerkraft-Auslaufs körnigen Gutes aus einem Bunker. 2.Teil", *Aufbereitungs-Technik*, vol. 26, pp. 130–135, 1985.

[FIR 86] FIREWICZ H., "Kinetik des Schwerkraft-Auslaufs körnigen Gutes aus einem Bunker. 3.Teil", *Aufbereitungs-Technik*, vol. 27, pp. 157–166, 1986.

[FIR 88] FIREWICZ H., "Kinetik des Schwerkraft-Auslaufs körnigen Gutes aus einem Bunker. 4.Teil", *Aufbereitungs-Technik*, vol. 29, pp. 61–70, 1988.

[FOS 84] FOSCOLO P.U., GILIBARO L.G., "A fully predictive criterion for the transition between particulate and aggregative fluidization", *Chemical Engineering Science*, vol. 39, p. 1667, 1984.

[FOS 87] FOSCOLO P.U., GILIBARO L.G., "Fluid dynamic stability of fluidised suspensions: the particle bed model", *Chemical Engineering Science*, vol. 42, p. 1489, 1987.

[GAB 70a] GABOR J.D., "Heat transfer to particle beds with gas flow less than or equal to that required for incipient fluidization", *Chemical Engineering Science*, vol. 25, p. 979, 1970.

[GAB 70b] GABOR J.D., "Wall-to-bed heat transfer in fluidized and packed beds", *Chemical Engineering Progress Symposium Series*, vol. 66, p. 76, 1970.

[GAN 08] GANESAN V., ROSENTRATER K.A., MUTHUKUMARAPPAN K., "Flowability and handling characteristics of bulks solids and powders. A review with implications for DDGS", *Biosystems Engineering*, vol. 101, pp. 425–435, 2008.

[GEL 72] GELDART D., "The effect of particle size and size distribution on the behaviour of gas-fluidized beds", *Powder Technology*, vol. 6, p. 201, 1972.

[GEL 73] GELDART D., "Types of gas fluidization", *Powder Technology*, vol. 7, p. 285, 1973.

[GEL 79] GELDART D., CULLINAN J., GEORGHIADES S. *et al.*, "The effect of fines on entrainement from gas fluidised beds", *Transactions of the Institution of Chemical Engineers*, vol. 57, p. 269, 1979.

[GEL 81] GELDART D., ABRAHAMSEN A.R., "Fluidization of fine porous powders", *Chemical Engineering Progress Symposium Series*, vol. 77, p. 160, 1981.

[GEL 86] GELDART D., *Gas Fluidization Technology*, John Wiley & Sons, 1986.

[GEN 68] GENETTI W.E., KNUDSEN J.G., "Heat transfer in a dilute-phase fluidised bed tubular heat exchanger", *Institution of Chemical Engineers Symposium Series (London)*, no. 30, p. 147, 1968.

[GIU 69] GIUNTA J.S., "Flow patterns of granular materials in flat-bottom bins", *Journal of Engineering for Industry (Transactions of the ASME)*, vol. 91, pp. 406–413, 1969.

[GRA 87] GRAHAM D.P., TAIT A.R., WADMORE R.S., "Measurement and prediction of flow patterns of granular solids in cylindrical vessels", *Powder Technology*, vol. 50, pp. 65–76, 1987.

[GUT 91] GUTFINGER C., CHEN W.H., "An approximate theory of fluidized-bed coating", *Chemical Engineering Progress Symposium Series*, vol. 66, no. 101, 1991.

[HAA 09] HAAR A., VON KARMAN T., "Zur Theorie des Spannungszstände in plastischen und sandartigen Medien", *Nachrichten von der Gesellschaft des Wissenschaften zu Goettingen, Mathematisch-Physikalische Klasse*, vol. 1909, pp. 201–208, 1909.

[HAA 83] HAAKER G., RADEMACHER F.J.C., "Direkte Messung der Fließeigenschaften von Schüttgütern mit einem abgeänderten Triaxial-Gerät", *Aufbereitungs-Technik*, no. 11, pp. 647–655, 1983.

[HAG 52] HAGEN G., "Druck und Bewegung des trockenen Sandes", *Berliner Monatsberichte. Akademic der Wissenschaft*, pp. 35–42, 1852.

[HAM 37] HAMAKER H.C., "The London-van der Waals attraction between spherical particules", *Physica IV*, no. 10, November 1937.

[HAN 74] HANCOCK A.W., NEDDERMAN R.M., "Prediction of stresses on vertical bunker walls", *Transactions of the Institution of Chemical Engineers*, vol. 52, pp. 170–179, 1974.

[HAR 61] HARRISON D., DAVIDSON J.F., KOCK J.W., "On the nature of aggregative and particulate fluidisation", *Transactions of the Institution of Chemical Engineers*, vol. 39, p. 202, 1961.

[HAR 87] HARNBY N., HAWKINS A.E., VANDAME D., "The use of bulk density determination as a means of typifying the flow characteristics of loosely compacted powders under conditions of variable relative humidity", *Chemical Engineering Science*, vol. 42, no. 4, pp. 879–888, 1987.

[HAR 02] HARI J.L., Réflexion sur des méthodes moins classiques et la préparation des poudres à la mesure, Societé Française de Génie des Procédés, Report, Compiègne, 2002.

[HAU 67] HAUSNER H.H., "Friction conditions in a mass of metal powder", *International Journal of Powder Metallurgy*, vol. 3, no. 4, pp. 7–13, 1967.

[HEC 61] HECKEL R.W., "Density pressure relationships in powder compaction", *Transactions of the Metallurgical Society of AIME*, vol. 261, p. 671, 1961.

[HER 81] HERTZ H., "Ueber die Berührung fester ekstischer Körper", *Journal für die reine und angewandte Mathematik*, vol. 92, pp. 156–171, 1881.

[HIE 84] HIESTAND H.E.N., SMITH D.P., "Indices of tableting performance", *Powder Technology*, vol. 38, pp. 145–159, 1984.

[HOL 83] HOLLEY C.A., "Agitative agglomeration methods", *International Powder and Bulk Solids Handling and Processing Conference*, Atlanta, pp. 99–108, 24 May 1983.

[HOP 80] HOPPE H., VON FISENHART-ROTHE M., "Druckverhältnisse bei der Silierung von Schüttgütern (Teil II und III)", *Aufbereitungs Technik*, no. 2, pp. 81–84, 1980.

[HUT 77] HUTTENRAUCH R., "The mechanism of tablet forming. A new conception", *First International Conference Pharm. Techn.*, Paris, vol. IV, 1977.

[ILA 02] ILARI J.L., "Flow properties of industrial dairy powders", *Le Lait*, vol. 82, pp. 383–399, 2002.

[ILE 08] ILELEJI K.E., ZHOU B., "The angle of repose of bulk corn stover particles", *Powder Technology*, vol. 187, pp. 110–118, 2008.

[ILS 96] ILSE M.F., GELDART D., "Characterising semi-cohesive powders using angle of repose", *Particle and Particle Systems Characterization*, vol. 13, pp. 254–259, 1996.

[JAE 89] JAEGER H.M., LIU C.H., NAGEL S.R., "Relaxation at the angle of repose", *Physical Review Letters*, vol. 62, no. 1, pp. 40–43, 1989.

[JAN 95] JANSSEN H.A., "Versuche über Getreidedruck in Silozellen", *Zeitschrift des vereines deutches Ingenieure*, vol. 39, pp. 1045–1049, 1895.

[JEN 59] JENIKE A.W., SHIELD R.T., "On the plastic flow of Coulomb solids beyond original failure", *Journal of Applied Mechanics*, vol. 26, pp. 599–602, 1959.

[JEN 60] JENIKE A.W., ELSEY P.J., WOOLLEY R.H., "Flow properties of bulk solids", *Proceedings of the American Society for Testing Materials*, vol. 60, pp. 1168–1181, 1960.

[JEN 62a] JENIKE A.W., "Gravity flow of solids", *Transactions of the Institution of Chemical Engineers*, vol. 40, pp. 264–271, 1962.

[JEN 62b] JENIKE A.W., "Slope stability in axial symmetry", *Proceedings of the 5th Symposium on Rocks Mechanics held at the University of Minnesota*, pp. 689–710, May 1962.

[JEN 64a] JENIKE A.W., "Entwicklung eines Verfahrens zur Verbesserung des Fliessverhaltens von gebunkerten Schüttgütern", *Bergbauwissenschaften*, vol. 11, no. 19, pp. 443–447, 1964.

[JEN 64b] JENIKE A.W., "Steady gravity flow of frictional-cohesive solids in converging channels", *Journal of Applied Mechanics*, vol. 31, pp. 5–11, 1964.

[JEN 65] JENIKE A.W., "Gravity flow of frictional-cohesive solids. Convergence to radial stress fields", *Journal of Applied Mechanics*, vol. 32, pp. 205–207, 1965.

[JEN 68] JENIKE A.W., JOHANSON J.R., "Bins loads", *Journal of the Structural Division*, vol. 94, no. 4, pp. 1011–1041, 1968.

[JEN 69a] JENIKE A.W., JOHANSON J.R., "Bin loads", *Journal of the Structural Division*, vol. 95, no. 11, pp. 2509–2510, 1969.

[JEN 69b] JENIKE A.W., JOHANSON J.R., "On the theory of bin loads", *Journal of Engineering for Industry*, vol. 91, pp. 339–344, 1969.

[JEN 71] JENIKE A.W., Johanson J.R., "Fliessgerechte Siloformen für Schüttgüter", *Aufbereitungs-Technik*, no. 6, pp. 309–317, 1971.

[JEN 87] JENIKE A.W., "A theory of flow of particulate solids in converging and diverging channels based on a conical yield function", *Powder Technology*, vol. 50, pp. 229–236, 1987.

[JOH 58] JOHNSON K.L., "A note on the adhesion of elastic solids", *British Journal of Applied Physics*, vol. 9, pp. 199–200, May 1958.

[JOH 64a] JOHANSON J.R., "Stress and velocity fields in the gravity flow of bulk solids", *Journal of Applied Mechanics*, vol. 31, pp. 499–506, 1964.

[JOH 64b] JOHANSON J.R., COLIJN H., "New design criteria for hoppers and bins", Iron and Steel Engineers, pp. 85–104, 1964.

[JOH 65] JOHANSON J.R., "Methods of calculating rate of discharge from hoppers and bins", *Transactions of the Society of Mining Engineers*, vol. 232, pp. 69–80, 1965.

[JOH 66a] JOHANSON J.R., "The use of flow-corrective inserts in bins", *Journal of Engineering for Industry*, pp. 224–230, May 1966.

[JOH 66b] JOHANSON J.R., KLEYSTEUBER W.K., "Flow corrective inserts in bins", *Chemical Engineering Progress*, vol. 62, no. 11, pp. 79–83, 1966.

[JOH 71] JOHNSON K.L., KENDALL K., ROBERTS A.D., "Surface energy and the contact of elastic solids", *Proceedings of the Royal Society of London A*, vol. 324, pp. 301–313, 1971.

[JOH 71/72] JOHANSON J.R., "Modeling flow of bulk solids", *Powder Technology*, vol. 5, pp. 93–99, 1971/72.

[JOH 79a] JOHANSON J.R., "Two-phase flow effects in solids processing and handling", *Chemical Engineering*, vol. 86, pp. 77–86, 1979.

[JOH 79b] JOHANSON J.R., COLIJN H., "New design criteria for hoppers and bins", *Chemical Engineering*, vol. 86, pp. 77–86, 1979.

[JUH 85] JUHASZ Z., "A method for characterization of bulk structure and bulk stability of powders", *Powder Technology*, vol. 42, pp. 123–129, 1985.

[KAW 71] KAWAKITA K., LÜDDE K.-H., "Some considerations on powder compression equations", *Powder Technology*, vol. 4, p. 61, 1971.

[KEH 71] KEHOE P.W.K., DAVIDSON J.F., "Continuously slugging fluidised beds", *Institution of Chemical Engineers Symposium Series*, vol. 331, p. 97, 1971.

[KEN 86] KENDALL K., "Inadequacy of Coulomb's friction law for particle assemblies", *Nature*, vol. 319, no. 6050, pp. 203–205, 1986.

[KIM 85] KIM Y., MACHIDA K., TAGA T. *et al.*, "Structure redetermination and packing analysis of aspirin crystals", *Chemical & Pharmaceutical Bulletin*, vol. 33, no. 7, p. 2641, 1985.

[KOU 01] KOUADRI-HENNI A., BENHASSAINE A., "Corrélations granularité coulabilité: étude d'un mélange bimodal d'une farine de blé tendre", *Récents Progrès en Génie des Procédés*, vol. 15, no. 77, pp. 109–114, 2001.

[KRU 67] KRUPICZKA R., "Analysis of thermal conductivity in granular materials", *International Chemical Engineering*, vol. 7, no. 1, p. 122, 1967.

[KRU 92] KRUYT N.P., VERËL W.J.Th., "Experimental and theoretical study of rapid flows of cohesionless granular materials down inclined chutes", *Powder Technology*, vol. 73, pp. 109–115, 1992.

[KUN 68] KUNII D., LEVENSPIEL O., "Bubbling bed model", *Industrial & Engineering Chemistry Fundamentals*, vol. 7, no. 3, p. 446, 1968.

[KUR 76] KURZ H.P., "Messungvon Schüttguteigenschaftenam Schergerätnach Jenike", *Verfahrenstechnik*, vol. 10, p. 68, 1976.

[LAK 75] LAKSHMAN RAO V., VENKATERWARLU D., "Internal pressures in flowing granular materials from mass flow hoppers", *Powder Technology*, vol. 11, pp. 133–146, 1975.

[LAN 94] LANGSTON P.A., TÜZÜN U., HEYES D.M., "Continuous potential discrete particle simulations of stress and velocity fields in hoppers: transition from fluid to granular flow", *Chemical Engineering Science*, vol. 49, no. 8, pp. 1259–1275, 1994.

[LAN 95] LANGSTON P.A., TÜZÜN U., HEYES D.M., "Discrete element simulation of granular flow in 2D and 3D hoppers: dependence of discharge rate and wall stress on particle interactions", *Chemical Engineering Science*, vol. 50, no. 6, pp. 967–987, 1995.

[LAR 74] LAROZE S., *Résistance des matériaux et structures*, Masson et Eyrolles, 1974.

[LAT 68] LATHAM R., HAMILTON C., POTTER O.E., "Back mixing and chemical reaction in fluidised beds", *British Chemical Engineering*, vol. 13, no. 5, p. 666, 1968.

[LEV 57] LEVA M., "Correlations in fluidized systems", *Chemical Engineering*, p. 266, November 1957.

[LIN 73] LINKSON P.B., GLASTONBURY J.R., DUFFY G.J., "The mechanism of granule growth in wet pelletising", *Transactions of the Institution of Chemical Engineers*, vol. 51, no. 3, pp. 251–259, 1973.

[LU 97] LU Z., NEGI S.C., JOFRIET J.C., "A numerical model for flow of granular materials in silos. Part 1: model development", *Journal of Agricultural Engineering Research*, vol. 68, pp. 223–229, 1997.

[LUO 93] LUONG M.P., "Flow characteristics of granular bulk materials", *Particle and Particle Systems Characterization*, vol. 10, pp. 79–85, 1993.

[LYK 79a] LYKLEMA J.W., "Computer simulations of a rough sphere fluid. Part I", *Physica*, vol. 96A, pp. 573–593, 1979.

[LYK 79b] LYKLEMA J.W., "Computer simulations of a rough sphere fluid. Part II. Comparison with stochastic models", *Physica*, vol. 96A, pp. 594–605, 1979.

[MAC 91] MAC ATEE K., BERMES S., GOLDBERG E., "Tivar-88 provides reliable gravity discharge from coal storage silos", *Bulletin of Solids Handling*, vol. 11, pp. 79–83, 1991.

[MAT 69] MATSEN J.M., HOVMAND S., DAVIDSON J.F., "Expansion of fluidized beds in slug flow", *Chemical Engineering Science*, vol. 24, p. 1743, 1969.

[MIC 84] MICHALOWSKI R.L., "Flow of granular material through a plane hopper", *Powder Technology*, vol. 39, pp. 29–40, 1984.

[MIN 49] MINDLIN R.D., "Compliance of elastic bodies in contact", *Journal of Applied Mechanics*, vol. 16, no. 3, pp. 259–268, 1949.

[MIN 53] MINDLIN R.D., DERESIEWICZ H., "Elastic spheres in contact under varying oblique forces", *Journal of Applied Mechanics*, vol. 20, no. 3, pp. 327–344, 1953.

[MOH 06] MOHR O., *Abhandlung auf dem Gebiet der technischen Mechanik*, Berlin, 1906.

[MUL 71] MULLIN J.W., *Crystallisation*, 2nd ed., Butterworths, 1971.

[MUL 72] MULLINS W.W., "Stochastic theory of particle flow under gravity", *Journal of Applied Physics*, vol. 43, pp. 665–678, 1972.

[MUR 82] MURTHY D.V.S., ANANTH M.S., "A one parameter model for granulation", *The Chemical Engineering Journal*, vol. 23, pp. 177–183, 1982.

[NED 83] NEDDERMAN R.M., TÜZÜN U., THORPE R.B., "The effect of interstitial air pressure gradients on the discharge from bins", *Powder Technology*, vol. 35, pp. 69–81, 1983.

[NED 92] NEDDERMAN R., *Statics and Kinematics of Granular Materials*, Cambridge University Press, 1992.

[NEM 80] NEMAT-NASSER S., "On behavior of granular materials in simple shear", *Soils and Foundations*, vol. 20, no. 3, pp. 59–63, 1980.

[NEW 58] NEWITH D.M., CONWAY-JONES J.M., "A contribution to the theory and practice of granulation", *Transactions of the Institution of Chemical Engineers*, vol. 36, p. 422, 1958.

[NGU 79] NGUYEN T.V., "Gravity flow of granular materials in conical hoppers", *Journal of Applied Mechanics*, vol. 46, pp. 529–535, 1979.

[NGU 80] NGUYEN T.V., BRENNEN C.E., SABERSKY R.H., "Funnel flow in hoppers", *Journal of Applied Mechanics*, vol. 47, pp. 729–735, 1980.

[NIC 62] NICKLIN D.J., "Two-phase bubble flow", *Chemical Engineering Science*, vol. 17, p. 693, 1962.

[NYQ 84] NYQVIST H., "Measurement of flow properties in large scale tablet productions", *International Journal of Pharmaceutical Technology and Product Manufacturers*, vol. 5, pp. 21–24, 1984.

[OOM 85] OOMS M., ROBERTS A.W., "The effect of surface roughness on the design and performance of gravity flow systems", *Proceedings of the Technical program/10th Annual Powder and Bulk solids Conférence*, Exhibition, Rosement, IL, pp. 7–9, May 1985.

[PIE 69] PIETSCH W., HOFFMAN E., RUMPF H., "Tensile strength of moist agglomerates", *Industrial Engineering & Chemistry Product Research and Development*, vol. 8, no. 1, p. 58, 1969.

[PIL 64] PILPEL N., "The flow properties of magnesia", *Journal of Pharmacy and Pharmacology*, vol. 16, pp. 705–716, 1964.

[PLA 74] PLANK F.W., "Bunkergestaltung", *Zement-Kalk-Gips*, vol. 27, pp. 271–277, 1974.

[POT 98] POTAPOV A.V., CAMPBELL C.S., "A fast model for the simulation of non-round particles", *Granular Matter*, vol. 1, pp. 9–14, 1998.

[PRI 78] PRITCHETT J.W., BLAKE T.R., GARG S.K., "A numerical model of gas fluidized beds", *AIChE Symposium Series*, vol. 74, pp. 134–148, 1978.

[RAN 52] RANZ W.E., "Friction and transfer coefficients for single particles and packed beds", *Chemical Engineering Progress*, vol. 48, p. 247, 1952.

[REB 77] REBERTUS D.W., SANDO K.M., "Molecular dynamics simulation of a fluid of hard spherocylinders", *The Journal of Chemical Physics*, vol. 67, no. 6, pp. 2585–2590, 1977.

[REE 78] REES J.E., "Biopharmaceutical implications of compaction and consolidation in the design of drug dosage forms", *Bollettino Chimico Farmaceutico*, vol. 117, p. 375, 1978.

[REI 71] REIMBERT M., REIMBERT A., *Silos, théorie et pratique*, Eyrolles, 1971.

[RES 66] RESNICK W., HELED Y., KLEIN A. *et al.*, "Effect of differential pressure on flow of granular solids through orifices", *Industrial and Engineering Chemistry Fundamentals*, vol. 5, no. 3, pp. 392–396, 1966.

[RIC 54] RICHARDSON J.F., ZAKI W.N., "Sedimentation and fluidization", *Part 1 Trans. Inst. Chem. Engrs.*, vol. 32, p.35, 1954.

[RID 69] RIDGWAY K., GILASBY J., ROSSER P.H., "The effect of crystal hardness at the wall of a tabletting die", *Journal of Pharmacy and Pharmacology*, vol. 21, pp. 245–295, 1969.

[ROB 91] ROBERTS R.J., ROWE R.C., "The relationship between modulus of elasticity of organic solids and their molecular structure", *Powder Technology*, vol. 65, p. 139, 1991.

[RÖC 05] RÖCK M., SCHWEDES J., "Investigation on the caking behavior of bulk solids. Macroscale experiments", *Powder Technology*, vol. 157, pp. 121–127, 2005.

[RON 95] RONG G.H., NEGI S.C., JOFRIET J.C., "Simulation of flow behavior of bulk solids in bins. Part 1: model development and validation", *Journal of Agricultural Research*, vol. 62, pp. 247–256, 1995.

[ROS 58] ROSCOE K.H., SCHOFIELD A.N., WROTH C.P., "On the yielding of soils", *Géothnique*, vol. 8, pp. 22–53, 1958.

[ROS 59] ROSE H.E., TANAKA T., "Rate of discharge of granular materials from bins and hoppers", *The Engineer*, vol. 208, pp. 465–469, 1959.

[ROW 73] ROWE P.N., WIDMER A.J., "Variation in shape with size of bubbles in fluidised beds", *Chemical Engineering Science*, vol. 28, p. 980, 1973.

[RUM 58] RUMPF H., "Grundlagen und methoden des granulierens", *Chemie Ingenieur Technik*, vol. 30, no. 3, p. 144, 1958.

[SAL 88] SALENÇON J., *Mécanique des milieux continus. I concepts généraux. II Elasticité. Milieux Curvilignes*, Ellipses, 1988.

[SAT 85] SATIJA S., FAN L.-S., "Characteristics of slugging regime and transition to turbulent regime for fluidized beds of large coarse particles", *AIChE Journal*, vol. 31, pp. 1554–1562, 1985.

[SAV 77] SAVKOOR A.R., BRIGGS G.A.D., "The effect of tangential force on the contact of elastic solids in adhesion", *Proceedings of the Royal Society of London A*, vol. 356, pp. 103–114, 1977.

[SAV 81] SAVAGE S.B., SAYED M., "Gravity flow of coarse cohesionless granular materials in conical hoppers", *Zeitschrift für angewandte Mathematik und Physik*, vol. 32, no. 2, pp. 125–143, 1981.

[SCH 75a] SCHUBERT H., "Tensile Strength of agglomerates", *Powder Technology*, vol. 11, p. 107, 1975.

[SCH 75b] SCHUBERT H., HERRMANN W., RUMPF H., "Deformation behaviour of agglomerates under tensile stress", *Powder Technology*, vol. 11, p. 121, 1975.

[SCH 81] SCHUBERT H., "Principles of agglomeration", *International Journal of Chemical Engineering*, vol. 21, no. 3, p. 363, 1981.

[SCH 86] SCHOFIELD C., "Recent progress in solids processing: recent research on the storage of particulate materials in hoppers", *Chemical Engineering Research Design*, vol. 64, pp. 89–90, 1986.

[SCH 90] SCHWEDES J., SCHULZE D., "Measurement of flow properties of bulk solids", *Powder Technology*, vol. 61, pp. 59–68, 1990.

[SCH 96] SCHWEDES J., "Measurement of flow properties of bulk solids", *Powder Technology*, vol. 88, pp. 285–290, 1996.

[SCH 98] SCHULZE D., "Die Charakterisierung von Schüttgütern für Silo-auslegung und Fließfähigkeistsuntersuchungen", *Aufbereitungs-Technik*, vol. 39, no. 2, pp. 47–57, 1998.

[SCH 00] SCHWEDES J., "Testers for measuring flow properties of particulate solids", *Powder Handling and Processing*, vol. 12, no. 4, pp. 337–354, 2000.

[SHI 68] SHINOHARA K., IDEMITSU Y., GOTOH K. *et al.*, "Mechanism of gravity flow of particles from a hopper", *Industrial and Engineering Chemistry Process Design and Development*, vol. 7, pp. 378–383, 1968.

[SIL 01] SILBERT L.E., ERTAS D., GREST G.S. *et al.*, "Granular flow down an inclined plane: bagnold scaling and rheology", *Physical Review E*, vol. 64, no. 5, pp. 1–14, 2001.

[SMI 29] SMITH W.O., FOOTE P.D., BUSANG P.F., "Packing of homogeneous spheres", *Physical Review*, vol. 34, p. 1271, 1929.

[SMI 82] SMITH K.J., ARKUN Y., LITTMAN H., "Studies on modelling and control of spouted bed reactors", *Chemical Engineering Science*, vol. 17, no. 4, pp. 567–579, 1982.

[SPI 05] SPILLMANN A., VON ROHR P.R., WEINEKÄTTER R., "Modeling the torque requirement of a blade stirrer in bulk solids", *Chemical Engineering and Technology*, vol. 28, pp. 741–746, 2005.

[SUM 77] SUMMERS M.P., ENEVER R.P., CARLESS J.E., "Influence of crystal form on tensile strength of compacts of pharmaceutical materials", *Journal of Pharmaceutical Sciences*, vol. 66, no. 8, p. 1172, 1977.

[SUT 73] SUTTON H.M., RICHMOND R.A., "How to improve powder storage and discharge in hoppers by aeration", *Process Engineering*, no. 9, pp. 82–85, 1973.

[SWI 83] SWINKELS F.B., WILKINSON D.S., ARZT E. *et al.*, "Mechanisms of hot isostatic pressing", *Acta Metallurgica*, vol. 31, no. 11, p. 1829, 1983.

[TAK 79] TAKAHASHI M., YANAI H., TANAKA T., "An approximate theory for dynamic pressure of solids on mass-flow bins", *Journal of Chemical Engineering*, vol. 12, no. 5, pp. 369–375, 1979.

[TAL 63] TALMOR E., BENENATI R.F., "Solids mixing and circulation in gas fluidized beds", *AIChE Journal*, vol. 9, p. 536, 1963.

[TER 86] TER BORG L., "Einfluss des Wandmaterials auf das Auslaufverhalten von Schüttgütern aus Silos", *Chemie-Ingenieur-Technik*, vol. 58, no. 7, pp. 588–590, 1986.

[THO 91a] THOMPSON P.A., GREST G.S., "Granular flow: friction and the dilatancy transitions", *Physical Review Letters*, vol. 67, pp. 1751–1754, 1991.

[THO 91b] THORNTON C., "Interparticle sliding in the presence of adhesion", *Journal of Physics D: Applied Physics*, vol. 24, no. 11, pp. 1942–1946, 1991.

[THO 93] THORNTON C., "On the relationship between the modules of particulate media and the surface energy of the constituent particles", *Journal of Physics D: Applied Physics*, vol. 26, pp. 1587–1591, 1993.

[TIN 92] TING J.M., "Computational laboratory for discrete element geomechanics", *Journal of Computing in Civil Engineering*, vol. 6, no. 2, pp. 129–146, 1992.

[TIN 93] TING J.M., ROWELL J.D., MEACHUM L., "Influence of particle shape on the strength of ellipse-shaped granular assemblage", *Proceedings of the Second International Conference on Discrete Elements Methods*, Cambridge, MA, pp. 215–225, 1993.

[TRE 68] TRESCA M.H., "Mémoire sur l'écoulement des corps solides", *Mémoires présentés par divers savants à l'Académie royale des Sciences*, vol. 18, pp. 733–799, 1868.

[TSI 93] TSINONTIDES S.C., JACKSON R., "The mechanics of gas fluidized beds with an interval of stable fluidization", *Journal of Fluid Mechanics*, vol. 255, p. 237, 1993.

[TSU 92] TSUJI Y., TANAKA T., ISHIDA T., "Lagrangian numerical simulation of plug flow of cohesionless particles in a horizontal pipe", *Powder Technology*, vol. 71, pp. 239–250, 1992.

[TÜZ 79] TÜZÜN U., NEDDERMAN R.M., "Experimental evidence supporting kinematic modeling of the flow of granular media in the absence of air drag", *Powder Technology*, vol. 24, pp. 257–266, 1979.

[TÜZ 82a] TÜZÜN U., HOULSBY G.T., NEDDERMAN R.M. *et al.*, "The flow of granular materials II. Velocity, distributions in slow flow", *Chemical Engineering Science*, vol. 37, no. 12, pp. 1691–1709, 1982.

[TÜZ 82b] TÜZÜN U., NEDDERMAN R.M., "An investigation of the flow boundary during steady state discharge from a funnel-flow bunker", *Powder Technology*, vol. 31, pp. 27–43, 1982.

[TÜZ 83] TÜZÜN U., NEDDERMAN R.M., "Gravity flow of granular materials round obstacles", *Bulk Solids Handling*, vol. 3, pp. 507–517, 1983.

[VER 67] VERLET L., "Computer experiments on classical fluids. Part I. Thermodynamical properties of Lennard–Jones molecules", *Physical Review*, vol. 159, pp. 98–103, 1967.

[VER 68] VERLET L., "Computer experiments on classical fluids. Part II. Equilibrium correlations functions", *Physical Review*, vol. 165, pp. 201–214, 1968.

[VON 13] VON MISES R., "Mechanik des festen Körper im plastisch-diformablen Zustant", *Nachrichten von der Gesellschaft der Wissenschaftlichen zu Göttingen, Mathematisch-Physikalische Klasse*, pp. 582–592, 1913.

[VON 77] VON EISENHART-ROTHE M., PESCHL I.A.S.Z., "Powder testing techniques for solving industrial problems", *Chemical Engineering (N-Y)*, vol. 84, no. 7, pp. 97–103, 1977.

[VRE 58] VREEDENBERG H.A., "Heat transfer between a fluidized bed and a horizontal tube", *Chemical Engineering Science*, vol. 9, p. 52, 1958.

[WAH 81] WAHL R., "Effect of upstream discharge on down-stream processing", *Chemical Engineering Progress*, vol. 77, no. 6, pp. 76–79, 1981.

[WAL 66] WALKER D.M., "An approximate theory for pressures and arching in hoppers", *Chemical Engineering Science*, vol. 21, pp. 975–997, 1966.

[WAL 73a] WALTERS J.K., "A theoretical analysis of stresses in silos with vertical walls", *Chemical Engineering Science*, vol. 28, pp. 13–21, 1973.

[WAL 73b] WALTERS J.K., "A theoretical analysis of stresses in axially-symmetric hoppers and bunkers", *Chemical Engineering Science*, vol. 28, pp. 779–789, 1973.

[WEN 82] WEN C.Y., CHEN L.H., "Fluidized bed freeboard phenomena: entrainment and elutriation", *AIChE Journal*, vol. 28, p. 117, 1982.

[WIE 75] WIEGHARDT K., "Experiments in granular flow", *Annual Review of Fluid Mechanics*, vol. 7, pp. 89–114, 1975.

[WIL 85] WILMS H., SCHWEDES J., "Analysis of the active stress field in hoppers", *Powder Technology*, vol. 42, pp. 15–25, 1985.

[YAN 76] YANG W.-C., "A criterion for fast fluidization", *Third International Conference on the pneumatic transport of solids in pipes*, University of Bath, England, 7–9 April 1976.

[YOR 80] YORK P., "Powder failure testing. Pharmaceutical applications", *International Journal of Pharmaceutics*, vol. 6, pp. 89–117, 1980.

[ZAB 66] ZABRODSKY S.S., *Hydrodynamics and Heat Transfer in Fluidized Beds*, M.I.T. Press, Cambridge, MA, 1966.

[ZAN 97] ZANG D., FODA M.A., "An instability mechanism for the sliding motion of finite depth of bulk granular materials", *Acta Mechanica*, vol. 121, pp. 1–19, 1997.

[ZEN 57a] ZENZ F.A., "How solid catalysts behave", *Petroleum Refiner*, vol. 36, no. 4, pp. 173–178, 1957.

[ZEN 57b] ZENZ F.A., "How to predict gravity flow rates", *Petroleum Refiner*, vol. 36, no. 10, pp. 162–170, 1957.

[ZEN 58] ZENZ F.A., WEIL N.A., "A theoretical–empirical approach to the mechanism of particle entrainment from fluidized beds", *AIChE Journal*, vol. 4, p. 472, 1958.

[ZEN 68] ZENZ F.A., "Bubble formation and grid design", *AIChE Symposium Series (London)*, vol. 30, p. 136, 1968.

[ZEN 77] ZENZ F.A., "How flow phenomena affect design of fluidized beds", *Chemical Engineering*, vol. 84, no. 27, p. 81, 1977.

[ZEN 78] ZENZ F.A., "The fluid mechanics of bubbling beds", *Fibonacci Quarterly*, vol. 16, p. 171, 1978.

[ZEN 83] ZENZ F.A., "Particulate solids. The third fluid phase in chemical engineering", *Chemical Engineering*, vol. November, p. 61, 1983.

Index